The Employee-Coach Handbook

Francie Van Wirkus

The Employee-Coach Handbook

ISBN: 978-0-578-88550-6 (ebook)
ISBN: 978-0-578-88772-2 (paperback)

Edited by Christina Gomez, Shelly Hankes,
Griffin Jones, and Jeff Shilling
Cover design by Katie Jankowski
Beautiful Photography by Steven M. Smith
Formatting by Polgarus Studio

Acknowledgements

Thank you, Jeff, my first and one of the best coach of coaches.
You reeled me in when I was reckless newborn,
and today I continue to learn from your brilliant mind.
Thank you to my editors for helping me make sense of my own head.
Thank you to my Lean-Agile tribe of friends for your expertise and support.
Let's celebrate this accomplishment, but not be complacent.
There is still much to learn.

Dedication

For Alan and Michael.
You are wonderful.
When the three of us collaborate,
precious, uncommon, human magic is in the air.

Contents

Introduction

Congratulations, you're an employee-coach! Oh boy. What did you get yourself into now? A really great experience that might not always feel great. If that's the case, you are in good company of those working to balance being an employee while also being a great coach. Oh, and by the way, be a change leader of some change, or even an all-out enterprise transformation. Doing this work well is entirely possible, even though it feels impossible.

Whether you coach a mindset, a methodology, a framework, or any combination of those, it's going to be a fantastic journey with fantastic challenges and fantastic opportunities. The employee-coach team is a great place to be for growth and life-changing experiences. If you are willing.

Two Parts to this Handbook

This work will focus on two main parts of employee-coaching: the team and the individual. You will learn how to establish and sustain a strong foundation for a team of employee-coaches, whether it is a team of two or twenty. You will also learn how to establish and sustain your personal, employee-coach foundation, so you can be strategic and resilient. Expect to encounter the word *strategy* over and again, because it is the path to success. Also, expect to see some repetition in the recommendations for the team and the individual. Even though some of the strategies are the same, the experience and journey of the team and individual will be very different. Humans are incredibly dynamic; a team of humans exponentially so. Be willing to slow down and focus and be open to what arises in each journey.

To keep the recommendations and approaches in this handbook clear, you won't see citations to studies and quotes from others to support them. The lists and basic nature of this handbook is what makes it, well, a handbook. This approach does not mean I want you to blindly trust what you find here; if you have questions or conflicting thoughts, research them, and talk with other thought leaders, coaches, and colleagues.

The people who will benefit the most from this book are new employee-coaches, established employee-coaches, the coach of employee-coaches, and management of employee-coach teams. If you already have an employee-coach team, it is probably time for a fresh look at your team, and how it operates.

As you read, expect to feel challenged on your current practices and most importantly, your mindset. When you do feel challenged, don't throw this book out. Instead, read on and let the challenge happen, reflect on it as a team and personally. (If you do all that and still want to throw out the book, good work! And, give me a call.) So, feel heartily encouraged to rise to the challenge, be curious about the friction of the advice you find, and use it to refresh your team and your personal practice.

The Focus and Intent of this Handbook

The intent of this handbook is to provide you with distilled thoughts and advice on the most critical activities and mindsets needed to be a successful employee-coach, and team of employee-coaches. Use this book as your work and your job aid, not just a collection of ideas.

The content is agnostic about methodologies and mindsets, focused solely on the employee-coach team and individual. The strategic approaches and recommendations here will even hold up to types of employee-coaching that do not yet exist at the time this book was published. So, if you feel like you're missing something in the *how* of your role as you read, you are correct. Employee-coaching is a big topic. Remember the narrow focus of this book is on the what: a strategic and resilient employee-coach team and individual employee-coach.

Expect to find guidance and recommended approaches to create a strong foundation for the team of coaches and the individual coach. Even with specific guidance, the outcomes will vary. A lot. Humans and the systems they work and live in are dynamic and complex, so there are no absolutes in employee-coaching. It bears repeating; not every challenge or opportunity has solutions and answers.

> Client: "Can't you just give me a list?"
>
> Coach: "What would you do with a list?"
>
> Client: "I would do the stuff on the list, so I can say I'm done, and get back to work."
>
> Coach: "I could give you a list, and you could do it, but that's not coaching."
>
> Client: "I'm really busy. Just make me a list!"

Ah, *the list*. Here it is, in your hands: several lists, from a coach! They are designed to be simple and approachable but won't give you all the answers. You will find mostly the *what* in this book, not the *how*. You will find what is needed to set up and sustain a strategic and resilient employee-coach team, and all the moving parts inside that work. *How* this work is done lives inside a broad spectrum of application that is best explored within your organization, and hopefully with an experienced coach of coaches. Be willing to explore the *how* for your team, and with your team. Take what is given, here, with a pragmatic mindset, and a willingness to explore what it means to your organization's desired change and your employee-coach team.

No matter your expertise in a mindset or methodology, how you coach people can really make or break your intentions. I've seen brilliant coaches destroy trust and credibility in a matter of moments because of poor coaching. Many have similar, memorable experiences. The employee-coach is in an especially precarious position, having closer, official ties to the organization that potentially extend for years of a career.

In a perfect world, the employee-coach is first sent to a formal coaching education and training program that has a post-training learning group. Then, a coach of coaches would continue with education and explanation in the field, along with observing the new coach in action and providing feedback. This process would take several months to get started, and a few years to become established. Most organizations don't give their employees or management this much time, for all the right and wrong reasons. Employee-coaches are often quickly shuffled into a newly formed team, sometimes re-organized into one, without having even applied for the job.

Some employee-coach teams are quickly assembled and do not get any formal coaching education or training. Instead, they received an adjacent training or certification for the methodology or mindset of the desired change. Knowledge, no matter how deep and fantastic, does not make a coach. Even the most multi-skilled, brilliant, certified person can't simply be a coach to what they know and learned. The journey to becoming a good (maybe someday great, or even fantastic) coach needs strategic education, training, mentoring, and a willingness to learn.

This book cannot replace the value of all those experiences described, but it can accelerate new employee-coaches far better than going it alone. If your coaching office is ready for the next level, recently went through another (!) reorganization, or version 3.0 of a transformation, this is book is a healthy, strategic, starting point for you and your team.

Being Willing and Curious is not for Everyone

If I've given you the impression that this book is for every employee-coach team, that is not the case. To build a strategic and resilient team, the employees and management must be *willing* to continually improve and grow. Most importantly, they must be willing to make the time to do these things, so it is part of the essence of the team: *A team like us does things like this.*

This book is not for every individual, either. It's for the willing and the curious (notice I didn't write uncertain). If you're not sure about the commitment, that is expected and normal. Building a personal vision and strategy so that you can reach your full potential takes far more time and effort than most expect. This book invites you in to ponder and consider the possibilities. *Be willing.*

Often you will read the phrase, *need to* and the word, *should*. These words can make this work sound absolute and rigid. Overall, the work of coaching coaches is far from absolute or rigid. Remember, the message in this book is about the *what*: build a strategic and resilient employee-coach team and an individual employee-coach practice. This work demands the rigor of focus and well-defined action. The bottom line: this book is advice, so it may sound bossy at times.

So, the content is here for you and your team to test, to use, or to inspire you to find a new and better way. There are always new and better ways to be an employee-coach. Continuous learning and improvement is expected.

I know you want to dive in, but having some additional context will improve the chances of you successfully consuming and using the contents of this book. There are many right ways to view and position an employee-coach team and the change they lead. The additional context provided here is the best foundation to fit the advice offered:

- You want a strategic and resilient team
- Being willing and curious is not for everyone
- Change defined
- The employee-coach defined
- Why have an employee-coach capability?
- Leaders, change leaders, and the management team defined

You Want a Strategic and Resilient Team

The work of constantly finding new ways to remain profitable and relevant and valued by the customer, while keeping an eye on the future, never stops. Strategic and resilient teams are strong contributors to the success of a business. This book assumes you want strategic and resilient teams. Most organizations will say they want and have them, yet the reality of their employee survey results usually reveals a lot of room for improvement. No amount of team knowledge or expertise can overcome the lack of strategic intent or operations of a team. No amount of pressure on management to improve how they *drive for (strategic!) results* will change the current state.

The designers of an organization might have a chart showing how the people are *aligned* or organized in a strategic way. Yet time and again, survey results and team retrospectives reveal lack of strategy in daily, program, and portfolio work and actions. Examples include conflicting priorities, starting and stopping work, too much work in process that doesn't connect, and chronic, last-minute management overrides.

Before this section changes into an indulgent rant, let's focus: A strategic team in the context here is not a team that *delivers* strategy but one that *operates with strategy*. A large part of this book is dedicated to recommendations and approaches on how to build and then operate with strategy, which requires focused, relentless work.

A strategic team, one that operates with strategy, isn't perfect, but it has higher engagement and far fewer problems like the ones previously listed. Not convinced? Invest some time to research countless studies from major survey companies and thought leaders on the topic, so these words are not merely my opinion but factual data.

Resilient teams have been researched for some time; there is a lot of sport and business team data available to support the case for team resilience. And, 2020 re-ignited interest in it for both teams and individuals, as a virus, and the reactions to it, disrupted most businesses and our personal lives. Hard times may draw more attention to the topic, yet true team resilience is beneficial in good times as well.

Team resilience is the ability for a team to be challenged or stretched out of their normal state, and then quickly return to a new normal state, in which they have grown from the challenge. It's a mindset, not a process. The team has a collective, continuous-learning mindset that they use to quickly sense and adapt to conditions and experiences. Team resilience is similar to a dedicated athlete's resilience: When injured, the athlete does the difficult work to heal their injury and learn from what happened to prevent getting the same injury again. Resilient athletes experience challenges like everyone else, *and* they see and work through them differently. A large part of this book is designed to provide you with recommendations and approaches to create the environment for team resilience.

> Team resilience is similar to a dedicated athlete's resilience: When injured, the athlete does the difficult work to heal their injury and learn from what happened to prevent getting the same injury again.

Change Defined

The word *change* will be used to refer to any change, transformation, or other work the employee-coach team supports. When referring to organizations, a transformation is a complete change in the organization's form and environment. Not all coach work is for a transformation, but most coach work is designed to effect change. Let's keep it simple, and call it *change*.

Change partners will refer to anyone outside the employee-coach team who is leading the change in the organization. They could be the change enablement team, the change delivery team, organizational change management (OCM) team, communities of practice, even a loosely organized group of stakeholders. The employee-coach team's and management's relationship with the change partners is critical and will arise in topics throughout this book.

Customer and Client Defined

When the topic is broad and high level, the word customer will be used. For example, you will read a lot about having a customer focus and customer value. This is broad and strategic in nature. The employee-coach team is customer focused and wants to deliver customer value.

When the topic is focused in on the people the employee-coach actually coaches, the word client will be used. The employee-coach has clients whom they coach and consult.

The Employee Coach Defined

This book uses the term *employee-coach* to emphasize distinctions from other types of coaches. The specific definition refers to a formal coaching role within an organization for the benefit of improving the performance of teams and management within that same organization, that is, an internal coach. The contrast is different from a coach-for-hire, or an external consultant who coaches: a similar set of skills, but without the responsibility of sticking around to make sure things actually work in the long run. The other important distinction is from managers who have organizational authority. Employee-coaches seldom have official organization authority over those they coach, and even if they did, would never use it. The final distinction is between an employee who is a coach (employee coach) and someone who coaches employees rather than management, or temporary staff (employee-coach). The distinction is that employee-coaches always share a common mission, and accordingly, common performance outcomes, with the people they coach.

> The contrast is different from a coach-for-hire, or an external consultant who coaches: a similar set of skills, but without the responsibility of sticking around to make sure things actually work in the long run.

Why have an Employee-Coach Capability?

This book assumes the employee-coach is an accepted and valued part of a greater change in the transformation strategy of an organization. If it is not, use these basic, strategic reasons to help your organization invest the money, time, and people in an employee-coach capability.

Employee-coaches can:

1. Be a lever to help elevate the change or transformation requested by management of organization. Employee-coaches can be present daily at the local level, and in personal interactions, where those wanting the change can't go with the same focus and frequency.

2. Help people within the organization move and grow toward the change by unlocking the potential within themselves and by extension, the organization itself. Trainers and consultants can offer ways of working and mindset and knowledge and expertise; this is teaching and mentoring. This work is important, but it doesn't address resistance and other dysfunctions of change. An employee-coach uses teaching and mentoring, along with different skills to go much deeper, to work through resistance and dysfunction, which are certain to happen.

3. Inspire people to connect their work and behaviors to the change, so that it can be sustained. The employee-coach offers often-uncomfortable *perspective* like no other person in the organization

> can: what is and is not working. Even in the most positive situations, change is difficult; people can easily get discouraged. By practicing daily encouragement and curiosity, the employee coach helps teams and management stay motivated and consequently more aware of blind spots and opportunities.

Just because an employee-coach capability is a proven way to effect change as described, doesn't mean it's going to be effective in your organization. Use the recommendations and approaches in this handbook to create and sustain a strategic and resilient employee-coach team, and to offer the best chance for the team's success.

Leaders, Change Leaders, and the Management Team

The topic of true leaders and change leaders is easily its own book. There are many opinions on the topic, and few get the definition and the supporting mindset and actions right. Even fewer well-intentioned people can actually succeed in the *practice* of being a leader or change leader. This section provides a proactive, bottom-line clarity, so that the recommendations and approaches shared later are equally clear.

> Managing is not leading, directing is not leading. Leaders persuade people to follow them, with eyes and heart on the future, despite the noise and chaos of the current state.

Managing is not leading, directing is not leading. Leaders persuade people to follow them, with eyes and heart on the future, despite the noise and chaos of the current state. Many organizations mistake taking initiative and being curious for being a leader. These are great qualities of anyone, employee, manager, or parent. We hear corporate messages and people say, "Everyone is a leader here." It may sound motivating, but it's not usually intended to encourage employees to persuade others to follow them with the future in mind.

> Many organizations mistake taking initiative and being curious for being a leader.

Change leaders practice personal advocacy and repeated connection to vision to motivate and influence others to think about change. They use empathy and curiosity to meet people where they are and invite them to join the journey to the future. They are comfortable being uncomfortable about not having all the answers about the future. A change leader does not need positional power to be successful. This work is not easy and faces a lot of heat. Most successful change leaders are persistent and resilient.

For the focus of this book, the positional roles that manage and direct the employee-coach team at any level will be called *the management team*. For example, the manager and director of the employee-coach team are the management team. *Management* also includes active stakeholders of the team. More on stakeholders in an upcoming section.

You won't see the words *leadership* or *boss* in this work, because they are so often misused in language and print. Save those words, along with the ones defined above, as a topic for your next Lean Coffee or employee-coach discussion.

So now you have the context of this important work in focus, so that you can consume, and then use the offerings in these pages:

- You want a strategic and resilient team, so that success is actually achieved and sustained.
- Understand that being willing and curious is not for everyone.
- A single definition of change, so that the messages here are not clouded in change vs. transformation vs. some other variant of future state.
- A clearly defined employee-coach role.
- Some of the important ways it is worth investing in the employee-coach role to help effect change in the organization.

- Clear definitions for management, leaders and change leaders, so that you can absorb the deeper meaning of the messages offered to you here.

As you read on, be willing to apply this context to what you think and experience while reading this book. Even if you do not agree with the idea of investing in an employee-coach capability for your organization. Consider the ideas here, without commitment or obligation. Also, notice the different ways seemingly simple definitions can create conflict with the current state of your organization, and maybe your own perspective. In other words, *be willing*.

Part I
The Employee-Coach Team

Part I provides the recommended strategies and approaches to build and sustain a strategic and resilient employee-coach team.

How to Begin

All's well that begins well. A successful employee-coach team demands it. Whether you are a newly formed or an established team, this section applies to you. Really. It's never too late to pause to adjust and improve an existing team. Don't wait for the next round of transformation or next reorganization to make the most of your team now. Work with the future in mind. *Be willing.*

The employee-coach team is often subject to all manner of pressures and challenges that a typical team does not experience. Employee-coach teams are often pulled in many directions and can easily spin into a chaos mode that is difficult to get out of or change. When a coach team has a strategic foundation, they are far better equipped to work through the challenges and become stronger over time. Ultimately, this is team resiliency.

To stay focused as you begin or refresh an established employee-coach team, see the work in three parts: build a connected management team, build a strategic employee-coach team, and operate with strategy. Each of these constructs are critical to building and sustaining a strategic an employee-coach team and are detailed in the upcoming chapters. Some of it will feel repetitive;

that's because both management and the team need this foundation, with slightly different content.

Build a Connected Management Team
Build a Strategic and Resilient Employee-Coach Team
Operate with Strategy

A short list, with a lot to learn within it. You will be provided all the information and strategy needed to do this work. And, all of this work is part of the journey, with the long run in mind. If you choose to skip this important work, you will miss the chance to reach your full potential as a team, and the financial consequences of it. The long-term success of a team happens when it becomes a true asset to the organization. It doesn't matter if other teams in your organization have never been created this way. This time is different because a successful team of change leaders demands more. Make the choice to make the management and employee-coach teams stronger than ever. *Be willing.*

Chapter 1
Build a Connected Management Team

Identify who can Commit to Manage the Team

The challenges of managing a team of change leaders are demanding and relentless. Although managing the actual team isn't always easy, the real challenge is in *leading change.* The organization's people and systems will push against management and the team in ways never imagined. How these challenges arise are different in each organization, but the need from management is always the same: relentlessly connecting people to the vision, alignment to strategy and work, and support for the team.

The employee-coach team and management vary in size; some have many managers supporting them, others have one manager, or, half of one (fragmented across more than one team). Regardless of how management of the team is organized, they need to be a connected team. Before you think it's not possible in your organization, please read on.

Who are the Management of an Employee-Coach Team?

- A fully committed people manager for hiring, talent, career development discussions, and performance reviews. Effective managers at any level lead their teams and manage the systems they are in.
- A fully committed director or VP-level role to work with stakeholders with access and authority and/or position to move or mitigate large roadblocks on behalf of the team.

- Manager of the change enablement/transformation office, if separate team from the coaching team.
- Other stakeholders, such as from the product, delivery, or operations areas.
- Executive sponsor(s) of the employee-coach team.
- Executive sponsor(s) of the change enablement/transformation office, if separate from the coaching team.

This section defines what is needed to establish a strategic foundation for the management of the employee-coach team. Managing a team of change leaders is heavy work: provide the team with the actionable vision, alignment, and support the team's needs to make a measurable difference the organization's performance. We are not talking about the kind of feel-good statements tucked away in a slide deck that only appears for division meetings. The coach team needs the kind that actively informs the choices of the employee-coach team and supports them through the unpredictable heat of leading change. Management needs to be change leaders, leading a team of change leaders. What could possibly go wrong?

Employee-coaches need committed management to do the time-consuming work of removing impediments for the team, meeting with stakeholders and clients, and to relentlessly connect them to the vision. Many employee-coach teams have a ghost–as in never seen or heard from–for a manager or director. Usually because these people have too much work on their plates to focus well on any one thing. Managing change leaders is a full-time job, not night school. So now what?

> Managing change leaders is a full-time job, not night school.

The message to repeatedly send to those who help decide on the commitment level of the team's management is that leading a team of employee-coaches is different and needs lots of attention. Remember the change leader context from earlier? In most organizations, the roles and expectations of managing a team of change leaders are unlike any other. This is not easy work. If you need to, go for the metrics, some of the most compelling ways to get buy-in. Likely, the employee-coach team is connected to a change of some kind, and the success of the change needs a strategic employee-coach team. There are likely OKRs (objectives and key results) inside the change work that connects well to the function of the employee-coach team. Use them to make the case for committed management.

If the best-case scenario of 100% dedication of management is simply not possible, aim for having it possible in three months. The executive sponsors of the work the employee-coach team is organized around may not realize this type of decision is what they bargained for. Have a clear message on the reason for the change and communicate it often with patient urgency. Make monthly inspections of your progress toward the goal(s) of 100% dedication, and then adapt as needed to ensure you're on target.

Think of strategy as a vessel for the team. Successfully leading change needs fully committed people to help build a vessel capable of moving into the wind. When the headwinds kick up, and they most certainly will, the employee-coach team will be able to flex and adapt in the challenges. Building the vessel starts with getting everyone together to work on alignment and strategy. The following section includes guidance on how to get people together to begin the journey.

Set Direction, Targets, and Strategy Using Strategic Events

Many books advocate for management to set direction, targets, and strategy but few tell you what that actually looks like. This section gives you the *what*. It may look like a very long list to you; this is the work of building a connected team. In this focused, no-nonsense handbook, every single item in every list is worth doing. The work can feel impossible and overwhelming at times, so it can be tempting to skip some of the steps. Cutting corners on strategy and leading change is a recipe for extended and intense hardship on all.

You Need Strategic Events (likely several) for Management and the Team

The management team *and* the employee-coach team will need separate, strategic events to connect and build strategy. Both groups must build a practice of collaborating to align and connect, so this initial event is the first of many for both groups.

The guidance and strategy in this section applies for both the management event and the team event.

Guidance for the Design of a Strategic Event

Planning and designing a strategic event can take twice as long as the actual event. There are exceptions on both sides of that estimate.

- Don't over plan, don't under plan. Find the middle ground of good enough.
- Define the purpose of the event.
- Define the intended outcomes.
- Based on the purpose and intent, you will know who to invite to the event. You will need to be a focused and nimble group.

Although not a secret event, generously exclude everyone else who thinks they need to be there but is not a member of the team or its management. If this sounds odd, review the first section of this chapter, and reaffirm who needs to be involved. This is *not* an informative event; everyone present must add value. No lurkers or others *just observing.* Here's more on the topic:

- If you over-include, you will dilute the purpose and intent of your event, and likely perpetuate organizational problems rather than leading change.
- Consider inviting strategic guests, such as stakeholders or executive sponsors for sections of the event.

- Prepare everyone well for the event, so they understand this is a strategic event to align change leaders. Give them the definition of a change leader for context.
- Schedule an executive sponsor to kick off and close the event or have some poignant part of it.
- Do not begin the event with logistics of the session. Save those for after you center the group.
- Plan to center everyone at the start of the event: begin with an opening that looks and feels different from others.
- Plan to have ground rules created by the group at the start. Examples:
  - Be present, fully focused, no multitasking.
  - What's in the room, stays in the room.
  - No one person dominates discussions. If this happens, hold each other accountable.
  - Breaks; when and how many. Adjust as needed.
- Make time for a retrospective *in the event.*
- Since you may have more than one event, use feedback from the retrospective to inform your next event.
- Keep the retrospective simple: What went well today? What could be better?
- End the event with a definitive close and a motivating message about the future.
- Consider the executive sponsor for part of this work.

Roles Needed for Strategic Management and Team Events

Role	Skills and Talents Needed
Event Designer	Designs event with strategic intent. Experience planning and designing events, and if virtual, knowledge of effective collaborative tools.
Executive Sponsor	Invites guests. Participates in event. Ability to lead change, connect the team and others to overall vision of the enterprise, and inspire them.
Third Party Event Facilitator	Outside person to facilitate event. Needs courage and credibility to be able to push the team if needed.
Follow Through Facilitator	Facilitates the follow through activities. Could be member of management team. Must have credibility to push the team and perseverance to sustain the follow through.

Mindset Needed for a Successful Event

- Focus on purpose and intent.
- Keep focused on the customer (yes, stakeholders can be customers too).
    - Consider personas or an empty chair in the room, or a sign in the room or in your virtual background about the customer.
- No one dominates the event. Protect attendees from distractions and diversions created by their own egos.
- Keep it real, not orchestrated.
- Make the event feel different, from the invite all the way to the follow through. Leading a team of employee-coaches is different work. Leading change is different work.
- Make containers or sections inside the event for brainstorming, collaboration, ranting, and listening.

- Brainstorming and ranting (hopefully not at the same time) do not last all day.
- Allow some variations in the flow of the event, but keep it closely connected to the intent.
- Sense when there is an impediment holding back the flow of the event, and make adjustments to improve it.

Desired Outcomes for Management Event

- MVP (minimum viable product) of coaching team vision and strategy.
- Agreement on a way to further refine the vision and strategy.
- Alignment on basic, foundational parts of the coaching office.
- Agreement on how to collaborate going forward.
- Clarity of management team roles and responsibilities.
- Working agreement*
- A shared experience to strengthen their team bond.
- A backlog of work for management to do with some sense of urgency.
- Agreement on how to collaborate around the backlog of work.
- Definition of an employee-coach in the organization.
- Understanding of the same coach role in the industry and greater market.
- Basic employee-coach job description.
- Decision on how employee-coach team will be organized.
- Outline of a service model for the rest of the organization to engage employee-coach services
- Define basic coach selection process.
- Create basic plan for new hires to have the best start possible.
- Report out to the executive sponsors, stakeholders, and coach team if it exists. A report out is a live event (may be recorded) that briefly shares the outcomes of a strategic event to a target audience, and usually invites feedback.

- Establish a basic feedback loop for the feedback from the report out, and a process for adjustments.
- Next two strategic alignment meetings are scheduled.

* Sample Management Team Working Agreement

A team working agreement contains simple guidelines and expectations on how the team will work and collaborate together. It is solely created by them and for them. A working agreement keeps the management team on common ground and also helps them move forward, faster. When making a working agreement, keep it simple and focused. Here is an example:

Macadamia Management Team Working Agreement

- Primary form of contact
- Core hours
- Response time
- How will we collaborate?
- What to do if you miss a meeting

Contact

Our primary way of contacting each other will be an instant message. If there is no response after one hour, we will use text. If there is no response to the text after one hour, we will group text or call the person's cell phone.

1. Collaboration app instant message.
2. Text cell number.
3. Group text.
4. Call cell number.

Core Hours

We are available for contact between the hours of 7:00am and 7:00pm (time zone).

Response Time

We aim to respond to any contact from this team within one hour.

Emergencies

Emergencies that cannot wait until the next business day are communicated via text or group text.

Collaborate

We are a remote team; we collaborate using video conferencing and use a secure space on our collaboration app.

Absence

We proactively communicate our absence to the team. We are responsible for learning what we missed, and our team will make it easy for us by using the secure space on the collaboration app.

Desired Outcomes for the Coach Team Event

- Connection to the purpose of their team.
- Some feeling of connection to each other and management.
- Understanding of the connection between the team's purpose, the change vision, and the organization's vision.
- Team understands how the strategic foundation guides team and management decisions and actions.
- The management team's alignment work output (the foundational parts of the coaching office) is visible to the team.
- Awareness of the different organizational systems the team interacts with.
- Established ownership of basic team functions and operations.
- Team has a definition of done for their primary work products and services.
- MVP of team and possibly individual success measures.
- Agreement on how the team will:
    - Make decisions
    - Collaborate*
    - Resolve conflict
    - Engage the outside world to learn and grow
    - Build its cross-functional capabilities and knowledge
    - Continuously improve their team
- Report out of outcomes to the executive sponsors, and stakeholders.
- Establish a basic feedback loop for any feedback from the report out and make any adjustments.
- Team is included in the refinement of the team vision and strategy.

*Sample Coach Team Working Agreement

Walnut Team Working Agreement

- Primary form of contact
- Core hours
- Response time
- How will we collaborate?
- What to do if you miss a meeting

Contact
Our primary way of contacting each other will be an instant message. If there is no response after one hour we will text, and if no answer after another hour, we will call the cell phone.

1. Collaboration app instant message.
2. Text cell number.
3. Call cell number.

Core Hours
We are a two-continent team. Our core collaboration hours are 8:00am - 11:00am (time zone).

Response Time
We will aim to respond to any contact from this team within one hour.

Emergencies
Emergencies that cannot wait until the next business day are communicated via text or group text.

Collaborate
We are a remote team; we collaborate using video conferencing on our collaboration app.

Absence
We proactively communicate our absence to the team. We are responsible for learning what we missed, and our team will make it easy for us by using the team's space on the collaboration app.

Strategic Event Follow Through

Some organizations are fantastic at these alignment events, but they lack basic follow through, putting alignment and the people involved at risk. Most organizations have a large amount of number one priorities, along with multiple disruptions and distractions that make staying focused very difficult.

So how will these groups be successful after the strategic event is over? Usually, with a little prompting from the facilitator, the participants will know exactly what to do next. Expect for them to want more time together. This is an encouraging sign that they are invested in their cause. Expect several work items to come out of the events that will need to be continually visible to management and the team.

The management team's follow through can begin immediately by practicing relentlessly connecting people to vision. Vision needs no apology, no backseat. Vision is king and al-ways will be. People need to hear vision and strategy many times in a year to even begin to internalize it and then connect it to their work. The vision has been adequately shared when all members of management and the team can explain simply and directly how his or her work contributes to the vision and how the next person downstream in the process contributes as well.

Immediate Next Steps to Ensure Quality Follow Through

- Administrative assistant or project assistant helps the Follow Through Facilitator with this work for both teams.
- Schedule a cadence for management team collaboration on calendars.
- Make the backlog of work visible for both teams, if it wasn't done during the event.
- Continue to let go of work that is not part of being a manager of an employee-coach team.

Collaborate to Establish Alignment and Connection

Now that you have the guidance for successful strategic events and follow through for the outcomes of them, it's time to focus on the specifics of the management team's alignment work. This work intends to build a connected management team.

When management is connected to the vision, strategy, and overall intent of the team, they are far more likely to be able to help the team succeed. No amount of good intentions or sheer will can overcome a poorly aligned or fragmented management team. They need to be in alignment on a short list of team strategies so that they are all steering the team in the same direction. Alignment prevents blind spots in areas where major damage to the coaching team or the organization can be caused. Alignment builds team resilience from day one. Why does an employee-coach team need resilience? To keep pace with change and other disruptors in the undulating work of supporting organizational change or a full transformation. A resilient team is not bogged down in decision making or stuck in politics when the need or opportunity to pivot arises. To be clear, alignment leaves room for innovation and growth. Imagine alignment as the guard rails for a wide road; there is protection from running off the road, and a lot of room to drive on the road.

Getting a management team connected takes time and is a perfect example of the philosophy, *slow down to go faster.* Whether a new employee-coach team is forming, or an existing team needs a refresh, getting the answers to these questions will establish the foundation for a connected management team.

Strategic Alignment Event for Management

This section aims to grow genuine management involvement in the employee-coach team's success, and the success of the change. So many employee-coach teams lack genuine, strategic, management involvement. For all sorts of organizational and personal root causes, management in these organizations doesn't have the personal conviction of the change itself and/or of the coach team. If your organization is one of the few that has managers who are actual change leaders with conviction and involvement, your manage-ment team still needs strategic alignment, or a refresh of it.

There are many activities and experiences to connect a management team. Consider and create what will work best for your management. Easy to say, hard to do. I offer a recommended design for this event; feel free and inspired to design your own. The most important activity is to do it together, not using any other proxy of method or people. Resist the temptation to cut corners on this experience, or worse yet, skip it all together. This is the foundation of the work the management and coach team will do in the future. This work is leading, not managing. It cannot be cone through email or instant message.

The team's management will need someone, or a few someones to support the actual event: management will need help to report out and make visible the work that was done and the outcomes of it. An outside-party facilitator, from another division or work area, or perhaps your coach of coaches, will yield best results. You will get some fresh ideas on how get the most out of your sessions, and facilitator neutrality to enhance the overall experience.

Strategic Questions Needing Answers

It is recommended to hold separate, strategic events: one for management and one for the team. You will receive guidance on both kinds of events, beginning with the management team.

The strategic management event should be designed to answer some very simple but important questions. Many organizations have a bias for the complex, so you may find objections to how simple this work appears. Stick with the intent of building a connected management team. Others may see the opportunity of who is gathered together and want to pile on other work. The discussions and decisions that were put off in the past is now looking for a home in your strategic event. Don't allow other work into this event, because building alignment and strategy needs complete focus. The management team needs generous space for creative discussion and collaboration. It needs to *feel* different, too.

> The management team needs generous space for creative discussion and collaboration.

If you find your management team can't hone in on this work, you probably have a different, number-one problem: *why do we get sidetracked from what we agreed was the most important thing?* You are stuck, and you need to get unstuck. It's a huge topic outside the scope of this handbook, and best in the hands of an experienced coach.

Now for the details: one section is focused on the management and team's vision and strategy to get there; the other section focuses on defining the employee-coach role in your organization. The experience and discussion to answer the questions is just as important as the actual answers. Here are the recommended questions:

Management and Coach Team Strategy

- Why do we have a coaching office?
- What do we do as management?
- How will we be successful?
- Who does what?
- What is our vision?
- What is our strategy?
- How does our vision and strategy connect to the portfolio(s) we are a part of?
- How does our vision and strategy connect to the enterprise vision and strategy?
- How will we relentlessly connect our team(s) to our vision and strategy?

- What help do we need from our stakeholders and executive sponsor?

Strategic Questions for Selecting and Hiring Employee-coaches

- What is an employee-coach, here?
- What is the job description?
- What does this role look like in our industry? In the greater market?
- Why do we think this role can help accomplish our mission?
- How will they be organized? For example, centralized, decentralized, hub/spoke.
- How will we select people for the team now?
- How will we ensure new hires have the best start possible?

Hint: this is your meeting agenda. This is your work. These are simple questions with sometimes a very complicated discussion (or series of them) needed to find the answers. Often, this work needs to be done over a few half day sessions. If the meetings are virtual, expect it to take longer. This isn't bad; it's simply the nature of a quality, virtual event. Quality takes time. Yet making the time to answer these basic but complicated questions is how dedicated management establish connection to the work they lead and to each other. In fact, it's so impactful, I recommend walking through these questions once a year as part of a management team retrospective. Management of an existing employee-coach office might feel they are beyond this kind of work. Since everything can be improved, position the work as a reflective exercise focused on inspecting and adapting. It's a very impactful way to build or maintain resilience in the management team and ultimately the coaching team.

All of the management team must attend all of the discussions. Alignment and connection cannot happen unless everyone is there. No proxies. Resist the urge, the pressure, to annotate this important work or push it off to a later date. Eventually is now. Successful teams of any kind need to spend time together reflecting, creating and collaborating on their future. Management must do their work first, and then invite and encourage the employee-coach team to do the follow through work at the team level. All's well that begins well. Let's move on to the details of team strategy work.

Chapter 2
Build a Strategic and Resilient Employee-Coach Team

Now that you've got a vision and strategy (or refreshed ones) and an aligned management team, it's time to form the employee-coach team. All's well that begins well with employee-coach teams, too. Priority one is to begin forming a strategic foundation and connection for the team. This work requires the team and management to slow down and establish themselves at a time when there is usually a lot of urgency to get started. The truth is, spending time team building *is* getting started, and, is the most urgent work.

Beginning with strategy and connection is a fantastic investment in the success of your employee-coach team. Most teams are formed with little connection or a thin veneer of it, and then they are off and running. This collection of highly talented people can do the off-and-running part very well. Problems happen when challenges arise; impediments or a fantastic opportunity. The team is left spinning on a decision or many decisions. High performing teams of today need to be resilient; able to adapt, and creatively face constraints. Select from any number of books, methodologies or training events in the market to validate this concept and learn more about it.

When there is a Reorganization

Sometimes, a coach team is formed or re-formed out of a reorganization, rather than choice. An established team could be moved and assigned new management, or the team and management could be new to each other. Or, any other combination of moving people and changing their relationship systems. Right or wrong, good or bad, reorganizations are some of the best opportunities to establish alignment and connection.

Any reorganized team and their management must work through the change that has impacted them before they can move on to alignment and strategy. This is especially true for an employee-coach team. Many organizational change and relationship systems experts advise this type of work builds trust, and strengthens the team as well as helps adoption of the reason(s) for the reorganization.

Priority one is to make the time and space for the team and management to process or work through the reorganization change(s) individually and together. This work takes time, and it is nonlinear. Working through the change of a newly formed team has a spectrum of activities, and yet there are a few that are commonly recommended for most reorganizations.

Remember, you want to begin well. Make the most of this time for a healthy, strategic start here and now. So many times, organizations are focused on the logistical and technology impacts of reorganizations that they miss the most important part: the impact on its people. When this important step is missed, the team and management will suffer later. They will have no strategy, alignment, or connection to support them as they work and make decisions in the uncertainty and chaos that are sure to be a part of a reorganization.

In reorganizations, management needs time to work through their own change before being expected to manage (and hopefully lead) their team. They need a chance to talk and think through the change with their own manager(s) and peers, so that they can effectively manage their reorganized team. Not all organizations give management the chance to do this important work. Request and advocate for this time, using the supportive construct of leading change. Giving management a chance to process change is a strategic, strengthening activity, not a sign of weakness. *Slow down to go faster* absolutely applies to even the most rushed reorganization.

> Giving management a chance to process change is a strategic, strengthening activity, not a sign of weakness. *Slow down to go faster* absolutely applies to even the most rushed reorganization.

Here some practical ideas for management to consider for working through their own change impacts:

- Reflect: use the practice of personal, curious, reflection to cut through the noise and chaos of change to understand what emotions and questions are present.
    - Make a list or draw images of what bothers you about this change and what encourages you about it. This work is similar to a *hopes/fears* or a *high dream/low dream* exercise.
    - Make a list of all the questions you have. Maybe they won't all get answered, at least not at first, but writing them will get them out of your head.
    - Make a list of the people you want to talk with and why. Note if you need to talk with them right now, in the next week, the next month, for example.
- Create inclusion and connection: meet with the person(s) you report to strictly to talk about the change; no logistics or updates. Have an open discussion with each other about the impacts of this change. Is an organization mapping exercise needed? If you don't know each other, this can be a challenge, but it is worth the effort to start the conversation.
- Create inclusion and connection: meet with peers who are also impacted by the change. Exchange questions and thoughts with each other. Perhaps you will find some early opportunities to learn from or help each other.
- Create a boundary: making time to process change can't go on forever. So, how long is long enough to give management a chance? Those of us who have been shoved through change without a breath would say that even a day to think it through and have a few

conversations would help. Better, would be a few days that include a weekend, or other time away from work. For example, if the announcement is (classically) made on Friday, provide time and space until Tuesday before asking them to face their teams. The reporting chain above the manager or director needs to be available and open for meetings and conversation.

- Leading change is never perfect, and always takes unexpected turns that can cause long-term disruption and possible disengagement.
- Management who are given the chance to find some solid ground of their own are more likely to successfully lead change and build resilience in themselves and the team.
- Prepared managers are comfortable with not having all the answers because they feel included and connected to the change.
- Management who have reflected first can be more authentic with leading through the change, rather than appearing as an excited veneer of purposeless energy.

Teams need a chance to process and work through the change as well. They need time and space to reflect personally and as a team, and to voice their thoughts and ideas about the reorganization. When teams don't get a chance to process change, there is far more spinning, more rumors, and less actual work accomplished. Similar to management, many of us have been on a team that was reorganized and then expected to simply start running. Team meetings are set up, management are introduced, and then the expectation is to GO. While many high performing teams can do this, there is usually fallout from not giving the team a chance to process the impacts of the change.

Here are some practical ideas for teams to work through the change of a reorganization:

- Create inclusion and connection: time alone as a team to discuss and exchange ideas, without management. Start with open discussion, rather than one tied to what is good, what is bad about this change.

- Create connection: time with management to discuss the change and ask questions.
- Create inclusion and visibility: sometimes, a *Festivus*, or airing of the grievances, is needed. Most people know life isn't perfect, but they need to be seen and be heard about how they have been impacted. Emotions need a place to be let out, or they will infect the team's work, management, and hold back forward progress. A team that needs this sort of interaction isn't bad, just stuck.
- Create a boundary: know when to say when with the *Festivus*. There is a specific time for this work, and then there is a time to acknowledge that not all of the problems can be fixed now. Sometimes, team members need assurance that the past is not the present.

In case you have not noticed, this is difficult and unpredictable work. The way things *should be* for a reorganization rarely manifests in reorganizations of any kind of complexity. Persevere anyway, with the long run in mind. All's well that begins well.

Hire with Strategy
Coach Interview Questions

Every organization likely has their own custom job interview strategy and tasks. If the employee-coach role is new to your organization, it's best to review this section ahead of any interviews. An employee-coach role is different in the interview process too. You are hiring someone who is potentially going to coach, train, and influence many other employees far beyond the impact of most other roles in the organization. No pressure! While hiring the right person cannot be guaranteed, adjusting the interview questions for coaching candidates is a great first step.

As with all interviews, there are important mechanics to help you find diverse, authentic, quality candidates who will best fit your search. Listen for pat answers, the kind that the interviewee knows you want to hear, compared to real experiences that come alive when the candidate speaks. You don't know if this person is authentic or not, so be ready to dig deeper if you do get a vanilla answer. Try a simple, *"Tell me more about that,"* or *"Can you expand that into a personal experience?"*

Experienced coaches usually present as neutral in interviews and in describing their experiences; they may even *say* that they are neutral on specific topics. This is usually a promising view into their coaching style. On the other hand, a candidate who shows up having a solution to every problem or opportunity (pattern matching) is more likely a coach with a big hammer looking for a nail. Candidates who show up in employee-coach interviews with big hammers are not always unwanted, but they do not usually fare well as an employee-coach.

The Interviewers

Who performs the interviews is critical to the employee-coach hiring and selection process. Let's take a closer look at how these people can be strategic.

The Recruiter

A recruiter may be the screener to weed out the obvious, which can be very helpful, or not at all! The recruiter needs to understand the uniqueness of the employee-coach team. Sending an email with a job description and selection criteria is not enough. Spend time face to face if you can, sharing the obvious with them, so they have a better sense of how to select people for this special team.

The Experienced Practitioner

Consider existing employee-coaches experienced practitioners for the interview process. The strategy of this interview is a peer-to-peer evaluation, and also to understand if the candidate would fit with the team and the work.

Beyond the obvious internal question, *"Can I work with this person?"* The experienced coach is charged with looking for nuance and truth in experiences, and also looking to see if the candidate is a good fit for the team. *"Can this person work with us?"*

Keep in mind that highly experienced employee-coaches still need preparation for interviewing candidates, so prepare your people well. And, not all highly experienced practitioners can be successful, voice-of-the-company interviewers.

The Manager

The manager who the coach will report to should be involved in the process, too. The strategy of this person being in the interview is to observe and evaluate for satisfying current needs with the future in mind. In addition to answering the internal question, *"Can I manage this person?"* and, *"Can this person grow with us?"* Management seeks to understand if the coach can successfully fit in the organization and with the team. They also honestly inform the candidate of the current state of the team, the change, and the types of challenges to be expected.

Employee-Coach Interview Questions and Strategy

Strategy: Provide a broad view of their experiences and to validate their resume, cover letter, and other data.

- Describe your (methodology/mindset/framework of choice/coaching) journey.
    - Some people may want to draw this for you; be open to their story.
- Do you have a professional coaching certification?
    - If yes, how do you use it?
- What were the big learning moments on your journey so far?

Strategy: Reveal Behaviors of Continuous Learners, and Areas of Interest

- What are you learning right now?
- Who do you follow in the (methodology/mindset/framework of choice/coaching) and why?
- How do you continuously learn and improve?

Strategy: Reveal Coach Maturity, Experience, Mindset

- Have you coached teams before?
- Tell me about your coaching style.
- How do you prepare people to be coached by you?
- How do you approach a new team, when you don't know how they feel about having a coach?
- How do you approach new management, when you don't know where they stand on your presence there?
- Tell me about a time when you influenced someone above you.
- Tell me about a time when it was difficult to be an objective coach. Did you pivot, persevere?
- Tell me about a time when you worked with multiple, strong personalities.
- What other methodologies or mindsets do you have experience in?
- What do you believe you need to be successful in this role?
- What would you say to someone who continually resists your coaching advice?

Even with the best interview questions, finding the right coaches for the team may take some iteration. Since humans are involved, there are some random situations that become the outliers for success or failure. The good news is that success once again depends upon strategy, mostly. For example, when the team and management find themselves in a split decision on hiring a candidate, using the teams' vision and strategy as a filter or a comparison will inform the decision.

Align the Team for Strategy and Resilience

Now that you have an aligned and strategic management team, and hired the right talent, it's time to gather everyone together and connect them to the vision and strategy. This work may feel repetitive. After all, you just got your management team through two or more (successful!) strategic events. It is repetitive for a reason. Your team of change leaders needs the experience of strategy work and the outcomes of it, too.

Quality in-person or virtual events take time. This section provides the *what*: the important, can't-miss intent and content needed for successful, strategic team events. *How* your team and management do this work is up to you.

And, if you really want to sustain a strategic and resilient team of change leaders, you will host these events annually, for management and the team. Your business and organization changes so fast, that often a year of work can feel like five years. Imagine a team of change leaders working from a vision and mission created just two years ago! Remember, all's well that begins well.

Coach Team Strategy Event(s)

- Refer to the section Guidance for the Design of a Strategic Event in Chapter 1
- Need solid, high-level design of strategic event: intent and outcomes
- Roles needed, including sponsors, what's needed for follow through
- Mindset for success
- Follow through activities

Strategic Team Event Framework

Kickoff: Connection

- Message from the executive sponsor(s). Logistics can come later. This sponsor shares why this event is happening, why now, and why this time together is important. Avoid overusing the word *excited.*

So many managers overstate how excited they are; at best that message appears fake.

- Facilitator introduction and any event logistics.
- Vision connection exercise:
    - Share employee-coach vision and strategy.
    - Be transparent: share related portfolio information, such as amount invested in the work, and the amount in overall the transformation and change portfolio.
    - If it exists, share the enablement/transformation office vision and strategy.
    - Share the enterprise vision and strategy and how it connects
    - Discussion about how these visions and strategies connect, and if there are any gaps.
    - Focus on employee-coach team vision and deep dive into it, such as a hopes and fears exercise. Then, what will the team do with what they discover about hopes and fears?

Collaborate to Gain Alignment and Connection

This section of decisions is large; for best quality outcomes, separate the work into two events, or more if needed.

- Introduction: Share *intent* of management team alignment questions, then share alignment questions and answers.
- Group discussion on alignment, using custom examples of when alignment matters.
- What is an employee-coach *here?* What do we do? What don't we do? What is uncertain?
- What organization systems are we in?
- How do these systems affect our definition of an employee-coach?
- How do these systems help or hurt our success as an employee-coach?
- What does management own?
- What does our team own?

- What do we as individuals own?
- Who is our customer and who are our stakeholders?
- What is our intake for work?
- What is our definition of done?
- How will we measure our success? Aim for lightweight, strategic measures.

Team Resilience: Connection, Inclusion, Visibility

- How will we resolve conflict?
- How will we quickly pivot our focus?
- How will we continuously learn and improve together and from each other?
- How will we build our cross-functional capabilities?
- How will we use experiments to create and innovate?
- How will we engage the outside world?
- Do we want to learn from other employee-coaches who have gone before us?
- Are there specific thought leaders we want to follow or even engage in a discussion?
- Do we want to share our journey in white papers, blogs, or conferences? (Even one that our compliance and security teams would approve for general public release.)

Immediate Next Steps to Ensure Quality Follow Through

- Administrative assistant or project assistant helps the Follow Through Event Facilitator with this work.
- Schedule cadence for team collaboration.
- Make the backlog of work visible, if it wasn't done during the event.
- Schedule any follow up strategy event(s).

Establish a Cross-functional Team

Cross-functional teams are diverse in talent and perspective, yet all working on a common goal. This doesn't mean that only one person on a team can have a specific talent or skill, but that the team *has what it takes* in talent and skill to achieve their goals. Such teams, with the right supportive environment, can quickly sense and respond to changes. They are better able to innovate, even when there are significant constraints. These teams can be considered a corporate asset. Research the expert advice on this rich topic. In the mean time, imagine the business value of a persistent, resilient, coach team that strategically leads and coaches to the change to the organization. You want this for a team of employee-coaches.

Where to begin? It takes strategy and team building to create and sustain a cross-functional team. A good starting point is your organization's vision and priorities, because most have some mention of growing knowledge across teams. Maybe even the lofty goal of becoming a learning organization. Connect the work of the team to these high level messages, and you have a foundation to collaborate and create activities and learning investment strategies.

What might a cross-functional team of employee-coaches look like? A team with…

- Deep experience, little experience
- Difference in coaching styles (gentle nudgers, bottom liners, for example)
- Experiences in specific industries (financial, manufacturing, sales, medical)
- Varied certifications in coaching methods and frameworks
- Experience with different kinds of *coachees* (individuals, teams, managers, executives, CIO, entrepreneur)
- Varied types of adjacent disciplines related to coaching
- Varied personal strengths. You can explore weaknesses, but focusing on personal strengths is far more effective and productive
- Different life interests

Even though their organization may support knowledge building across teams, your clients may not be used to pair or shadow coaching work. Many organizational systems say they want knowledge sharing, but the environments their people work in don't usually support it. The coach team and management need to keep focused on the client's perspective when making decisions about pair and shadow work. All's well that begins well, so make the time to intentionally prepare your client for the experience of on-the-job learning. If your organization has messages about up-skilling its people or becoming a learning organization, make the connection for your client. Pair and shadow work is exactly the type of activity that fits into those organizational goals. An upcoming section, Words Make it Real: Words for Your Client, offers ideas on what you might say to a client to help them understand the value of pair or shadow work. Your words and actions will be encouraging for your clients to try it, and, you will be modeling a continuous learning and improvement behavior that they may want to test in their area.

Teams don't magically become cross-functional. It takes intentional effort by the team and management, and time. This work builds invaluable team knowledge and it also helps team members better understand their strengths and weaknesses as a team, and as individuals. Many who have gone before you have published different exercises, events, and experiences to help a team understand its strengths and weaknesses. With so many differ-ent options and variations, it's easy to get overwhelmed or dig in too deep. Try not to overcomplicate it.

Get the team involved in deciding the experiences they want and have them facilitate the events. The steps described here are the very beginning of the journey of your team becoming cross-functional. Who is strong in facilitating or leading events? Who has never done it and wants to learn? Pairing the seasoned facilitator with those willing to learn more can be a powerful experience for all.

Cross-Functional Team Building Exercises

Many of us have been on a team when an urgent email request came through for everyone to load their skills and talents in a team spreadsheet by the end of the week. I request you put down that spreadsheet and think about being human. Besides, the spreadsheet list of skills has been done before; it felt good to make it, and then what happened to it? Likely not much, or, it got into the wrong hands with the wrong intent and caused chaos. A spreadsheet of skills and certifications is not a very useful way to measure a team of humans. A better approach to measure a team is by looking at how they work together, and the value they deliver. This work can be translated to a picture or image to share with your investors and sponsors. Whatever exercises and events the team does to understand itself and the people who comprise it, resist the urge to make a spreadsheet, or an inventory of team skills and talents. The following section offers some cross-functional team growth ideas designed to inspire you to more strategic activities.

Cross-Functional Team Growth

Try any of these ideas to begin to grow or keep growing your team's cross-functional capabilities. Keep in mind teammates who are sensitive to sharing detailed personal information; these exercises don't need to go deep to meet the objective. Also, there are far more ways to grow as a team than listed here. The important thing is to give it a try, and not give up if one idea doesn't work for your team.

- Exercise to evaluate strengths. Many surveys are available to buy.
- Create trading cards that have the skills and talent details the team desires (example strengths, interests, what you love about coaching, what challenges you about coaching, personal fun facts). Keep them positive.
- Share resumes and one-pagers. Meet and talk about them together.
- Create a map of individual coaching journey on a giant sticky note. Call it a journey, a canvas, a page, whatever feels right for the team.

- Create a letter-shaped skill (such as T-shaped) canvas/page/poster for each person, with a work focus.

For an impactful, lasting experience with these team building exercises, use the following approach:

- Make the giant sticky notes visible by creating a gallery.
- Do a gallery walk with the goal to observe with an open mind and discuss. No judgment.
- Meet to discuss them, and then leave them visible for two weeks.
- Virtual and distributed teams can get creative with how to share and then make the journeys visible.
- Repeat the exercise, or some variation of it the next year. Or, if the team changed dramatically, don't wait a year.

Words Make it Real: The Team's Words

It takes persistence to keep learning, and even more to learn as a team. The employee-coach team must constantly challenge itself to keep growing. The team's management can help support and encourage their growth, but it's really up to the team to advocate for their future.

Keep it simple and effective by asking questions like the ones below:

- How can we get stronger at this?
- Who would like to pair with me?
- Who would like to shadow me?
- Is this a growth opportunity for anyone?
- Who has strength in this?
- This sounds like something that needs all of our superpowers.

As always, make the questions visible. The work will not become real unless they keep coming into the view of the team. Maybe the questions cannot be visible all the time for virtual teams, but they can be made visible at strategic times. Most importantly, practice using them, especially for team events like these:

- Team retrospective
- Team strategy and planning events
- Management and executive sponsor strategy events
- Corporate and/or methodology surveys

Words Make it Real: Words for Your Clients

Your client relationship deserves an explanation for the way the coach team will work with them, and what the experience might be like. Pairing, shadowing or any other combination will be most welcomed when the client is prepared for it in advance. Strategic and resilient coach teams state their path, and then make the journey. These potential statements can help you begin the conversation in a way that doesn't alarm the client or make anyone from the coach team look disorganized or incompetent. The statements contain intent, strategy, and assurance, because all's well that begins well.

Example 1
Our team's work to grow and continuously learn means that sometimes we work in pairs or shadow each other. Know that you'll get the same confidentiality and great service. Similar to a teaching hospital, we have growing coaches who join experienced coaches in the work, so that they can accelerate their learning, and the change.

Example 2
Part of our team's focus to grow ourselves along with the organization includes our coaches pairing on work or shadowing each other. We can accelerate our learning and expertise when we share it with our teammates. Expect the same professional service and confidentiality as if it were only me as your coach. You'll also benefit from the perspective of another coaching mind on your work, which will help shape your overall experience.

Chapter 3
Operate with Strategy

Operate your coaching team and your personal coaching practice with strategy. Easy to say, hard to do, especially if coaching is a new concept in your organization. Employee-coach teams encounter a lot of turbulence; having strategy in which to operate creates an environment for success and builds resilience.

A business has value-add and non-value-add activities and invests in and innovates ways to keep adding value for its customers, so it can remain relevant and profitable. At the highest level, non-value-add team activities include work that is required of the team but does not add value to the finished product of *being* an employee-coach team. Your clients don't care that you have this work. An example is completing a project or time report each week. Value-add activities contribute to what the customer wants and is willing to pay for: the employee-coach presence and activities.

The important point here is that while the management, stakeholders, and sponsors of the employee-coach team are critical to the team's success, the team is the place where actual value is delivered. Everyone else has a role in supporting the team.

Operating with strategy is relentless work, if you do it right. Once you establish your employee-coach team, maintaining it with customer value at the forefront requires constant effort to sense and adapt. It's in the daily, weekly, and monthly habits of the team that enable the team to operate with strategy. It's in the full spectrum of daily, weekly, and monthly engagement and support of the management, stakeholders and sponsors that creates the environment for the team to make it happen.

If you've already established your employee-coach team, there is always time to refresh your practices and structure. Don't wait until year-end to make changes to something you already know could be better. Your team likely has more time and people available to help now than at the end of the year. Most all-hands meetings in December turn out to be few-hands meetings. Some established teams are more comfortable with a two or three-day refresh of their entire employee-coach business. Other teams; make an incremental change roadmap for the parts of their business they want to inspect and adapt over a monthly cadence. In both examples, it takes time, perseverance, and follow through; difficult work, even for a team of talented employee-coaches.

This section highlights some of the best ways to get your team to begin to operate with strategy.

Build a Strategic Foundation

Vision is king. A successful business needs to know where it's going and why. It's worth repeating that a team vision and strategy provide a strategic foundation for the team. And, management must relentlessly connect them to it and the enterprise vision and strategy.

Engage Investors

Team management, executive sponsors, and stakeholders are the investors in the employee-coach team. They make or help make the decisions on how the employee-coach capability is funded, success measures, and they help remove impediments affecting the team.

A business with investors who are aligned on vision, strategy and other basic aspects of the work is far more successful than when they disagree, or worse, are at war with each other. Investors who are aligned spend more time focusing on helping the team deliver customer value, and when there are challenges, they use their vision and strategy to find their way. Not only do the investors need to relate to each other, then must constantly connect with the team for three main reasons:

First, management must relentlessly connect the team to the vision of the coaching capability, to make certain they are always aiming for the target. No need to apologize for the vision again. The people who deliver business value need to hear over and over again where they are going and why. When people feel more connected to their work, they are more likely to be engaged, and care about outcomes.

Second, the team needs to hear continual business context, which is ever-changing in most organizations, so they are not speculating on their aim. The enterprise is full of organizations that often have conflicting interests, which can create a lot of turbulence for a coaching team. Even if the team has a focus in a specific area, it's not enough to shield them from the challenges of being in a larger organization. A team is more resilient to change and able to adjust and adapt when they continually hear about the ever-changing business context.

Third, the team needs its investors to remove impediments that they cannot remove themselves. The same concept applies to big opportunities; teams need their investors to promote and sponsor innovations that they can't do on their own. Successful teams know how to evaluate these problems and opportunities and determine which ones they need help with. When needed, engaged (angel?) investors are more than happy to help the team, because they have been strategically aligned.

Team Administration

Depending on team size and the type of employee-coaching, the team may need someone to help manage the administration and logistics of the team. What makes the most sense is to team up with the change team (enables or leads the transformation or change). If the change team has an experienced business manager in the role, it's possible this person could manage both teams. A better alternative would be for each team to have their own business manager, and then these two people work closely together.

For a strategic, resilient employee-coach team, the role of business or project manager extends well beyond reports and data collection. The intent of this role for an employee-coach team is organization of the work, the financials, and other team data. The role also represents the team with a customer focus when interacting with the rest of the organization. The employee-coach team is involved in some of this work, but they can't deliver coaching and keep up with the administration of themselves.

One obvious example of keeping the team organized is managing the team intake (discussed in greater detail later in this chapter) and the flow of work. The team developed their intake and established how work will flow, but they need help staying on track. Everyone on the team needs to use the system, and they need to communicate and collaborate to make it work. Leading change can be distracting work, even for the best of coaches. The business manager role can focus on the team intake and flow, and challenge the team to follow it and continuously improve it.

The business manager may also need to help the team's management stay focused on the work and the processes. Too many times, a manager is pressured to do a side project or one-off work, or to ask a team member to do this work. The business manager can help the team and management make sure all work goes through the rigors of the intake system. And, when there are problems or opportunities as the work flows through the team, it is made visible to everyone involved.

Employee-coach teams and change teams are often challenged to prove their value far more than other parts of the organization, and the business manager is often in the hot seat for this work. Leading change can be very unpopular with other parts of the company or with select senior management, who for any number of reasons doesn't agree with the work. Other parts of the organization may not even know about the change or the employee-coach team, until there are financial or other strategic discussions where the business manager is present. The business manager will face many challenges and needs to be resilient for the *"Who are you?"* moments that will likely occur over and again. This person needs the chops to ride the wave of challenges and uncertainty that accompanies change and a team of change management and have a willingness to inspect and adapt.

Project Assistant Support

When management and members of the employee-coach team take a close look at the work, they usually realize there are a lot of moving parts, and they also need administrative support. Some employee teams work with licensed content and using it within the good graces of the governing entity is nearly a full-time job. Other teams may need to navigate government regulations or global business guidelines. The project assistant can help establish and organize the team for this work and then maintain it for them if needed.

Having a project assistant also helps the team with *access*. Booking large events, whether in person or virtual, or getting on an executive's calendar are basic examples of the kind of magic this person can work for a team. Some virtual events need a producer, or someone to manage other parts of the event, so that the facilitator can focus on a quality event.

It is possible to share a project assistant with another team. If your organization is using licensed material that needs reporting and management, you may need a full-time project assistant, at least at the beginning. To succeed, the person in this role needs to be detail-oriented and have a mindset of perseverance and curiosity. They will also need to have an inspect and adapt mindset to fully support the team.

Marketing Activities and Assets

Early in the formation or refresh of the employee-coach team, decide what marketing and promotion the team needs to operate with strategy. Getting started doesn't mean perfection, so beware of spending too much time on refining your first assets. The goal is to get your team to the *market* within the organization, to establish presence and credibility.

The following questions can guide your team discussion and strategy for planning marketing activities and creating assets. Make time to meet together to answer them or revisit them if you've done this work in the past. You may be thinking, *"Oh brother, another event!"* Yes, strategic, resilient teams need continual care and collaboration.

Basic Marketing Strategy Questions

- We know why we are here, what we do, and who are customer is...
- How will we share who we are and our purpose (value!) with the organization? With clients?
- How will management provide favorable introductions of coach to client?
- How will we share our successes and opportunities?
- Will we need or want to share client testimonies?
- Who else might we partner with to accomplish our marketing goals?
- Are we part of any other team's marketing goals?

Examine the Enterprise for Other Teams like Yours

New teams forming around any work are often surprised to find a duplicate effort, or work that has very similar purpose, happening somewhere else in the enterprise. In transformations, impatient management wanting coaching for their area *right now* sometimes go off-road, outside the transformation enablement, to get what they want. This situation could be good or bad, and certainly warrants a strategic conversation about the overall change in the enterprise. In focus for this handbook, it's important the employee-coach team and its management are aware of all activity related to their work.

A newly formed coaching team and its management, or an established team can reflect by asking these questions:

Questions to Examine the Enterprise

- Are there other people like us in the enterprise, contracted or employee?
- Where are these people?
- Who is their executive sponsor?
- Are there other people in coaching roles, different from us, whether employee or vendor?
- Who is their executive sponsor?

Identify and Engage Business Partners

Operating the employee-coach team with strategy works best when you know who your business partners are, so you can establish and build relationships with them. This is valuable work for the coaching team and management to engage from their first day as a team, and thereafter. For already established employee-coach teams, taking a fresh look at your business partners each year is strategic work, more often if the enterprise had a reorganization.

Examples of Business Partners

- Training and development groups
- Building and facilities management
- Corporate communications
- Organizational change advisory
- Administrative professionals
- Audio visual

Business Partner Strategy Questions

What does it mean to review your business partners? Beyond making a list, the team needs to collaborate with management on growing and maintaining these relationships. Begin where you are. Make a list, then have the initial discuss to explore these topics:

- Who are our business partners?
- Who do we need and why?
- Who needs us and why?
- Are there any uncommon partners who are worth exploring?
- Who from the coach team or management can help grow and/or sustain this relationship?
- How can the coach team and management continually encourage their business partners to hold a systems thinking viewpoint?

Create the Team Intake System

Employee-coach teams that serve an enterprise will have a better chance at success with a formal intake system, so that the entire team has full visibility into the work. So many times, employee-coaches or their management are faced with a variety of unplanned work. From special favors to emergency work, employee-coaches and their management are easily pulled in many directions. Having a formal, visible intake system with a way to prioritize and order the work sets up the team's environment for success.

Entire books have been written about the tools and behaviors a team can use to manage the flow of its work for best outcomes. Kanban boards, program boards, Scrum backlog, product backlog, and other methods are options for an employee-coach team. As a team, with an inspect and adapt mindset, do the research, discuss the findings, and decide what to use. This book won't recommend specific tools; rather, encourage a strategy to keep the team's work visible, and to keep the team connected and included in the flow of work through the team. This guidance sounds very easy to do and yet it can be very challenging for any team to uphold it. As discussed earlier, the teams business manager can serve the team well by managing this work.

Intake System Guidance

Visual Management Guidance	Real World Interpretation
The work is readily viewable.	Our work is easily seen by the team and the people we work with.
We know our current truth.	Our board is a timely representation of reality.
We can see the flow our work from start to finish.	We can see all the useful steps, and the impediments to flow.
The team can see its capacity.	As a team, we know when we can pull more work.

One of the most common mistakes teams make with visual management is not putting enough effort and rigor into using their board(s), and then declare that the process doesn't work for them. These teams find truly amazing rationale on why they believe they are so different from the rest of the world that visual management doesn't fit them. The business manager and the team give the process time and use all the rigor recommended to make an honest experiment. Use the selected board and the activities recommended around it for at least one month. Give it an honest effort; after all, you're coaches, and you know how to build and sustain a behavior. After that time, if your team cannot execute on the processes supporting the board, inspect and discuss what happened, then start another experiment.

Prioritize and Order Work

Operating a coaching team with strategy demands that the work requested of the team be prioritized and ordered by the team *and* its management. Remember, the team is where the actual value is delivered.

While it's tempting for the team's management to make work priority and ordering decisions, push past the pressure in the name of a strategic, resilient team. Organizations often organize and pressure management to make work intake decisions without involving the team, because it's *perceived* as more efficient. Any time spared at the beginning is squandered by both the manager and the team, adjusting and compensating for uninformed, un-strategic decisions. In the long run, your team has a far better chance of working on the right thing at the right time when they are involved in the process from the beginning. Your team is cross-functional, and a fantastic source of truth on the best ways to deliver customer value.

The priority of the work isn't always the *order* of the work. For example, sometimes, work has to be done ahead of the top priority, to enable or pave the way. There are lightweight approaches available to help you collaborate and make the work visible.

Limit Work in Process (WIP)

So, the team has connected and aligned management, a vision, strategy, a body of work, and an intake system; all of these things help the team filter the right work at the right time. Limiting *how much work* the team is doing is another way to ensure the team is strategic. Work in process or progress, fondly called WIP, is the amount of work a team is working on at given time. When you have too much work, you are not likely to get much of anything done, and very likely to make poor decisions on the work you will complete. When a team limits the amount of work they have in flight, they can operate at a sustainable pace, and have better focus and can be more resilient to unplanned challenges. Rooted in Lean, the concept has been around forever, and applies to each of us personally, too. A great read on the topic by pragmatic experts: Personal Kanban: Mapping Work | Navigating Life by Jim Benson and Tonianne DeMaria Barry. The Agile Bettys podcast offers another way to learn about using Kanban in your personal life.

The team and management collaborate to define their WIP. This work usually includes several small experiments over time, and the results incrementally inform the team. While there are no absolutes, employee-coach work may have some patterns and trends that the team can learn from over time. Many organizations pressure employee-coach teams to be as predictable as other non-coach teams; they probably don't understand the work, which is something that can be resolved by the team's management or stakeholders Once again, having aligned management and stakeholders helps ground these conversations in value and customer focus.

On the word experiment: not everyone likes it. For a spectrum of reasons, the E-word can cause a lot of discomfort, most of it about wanting and needing to be predictable, and not wanting to be wrong. Others don't like to think of any work involving humans as an experiment. Don't worry. You don't need to use that word if it's going to cause problems. Focus on the construct of continuous learning and improvement.

The bottom line for all successful employee-coach teams: make time for continuous learning and improvement and slack time for other activity, such as team collaboration, reporting (gulp!) time, vacations/paid time off, and regular team and organization meetings. There are so many different expressions of making time for learning in organizations; the team can get really creative with ideas. For example, some offer eight hours of learning time a month, others offer a dollar amount to invest in the learning of their choice (think: conferences, certifications, learning events).

The greatest challenge any team faces is not getting the investment of money for the effort but in *making the time* for continuous learning. Relentlessly bake, weave, integrate, and negotiate learning into your strategy, and into the team's backlog of work. Specifically, limit the team's WIP to make the time. The team and management must constantly work to make continuous learning a priority. Resist even the most compelling rationale to skip the learning in favor of delivering more work. The effort to limit WIP may stumble a few times before it becomes part of the team's way of working, so persevere and use team reflection to inspect and adapt for the future. Be willing to care about the future, and the skills needed to excel in it.

> The greatest challenge any team faces is not getting the investment of money for the effort but in *making the time* for continuous learning.

Calling it Done

So you have aligned management, a vision and strategy, a body of work, an intake, and a way to limit the amount of work in flight. There is another activity and outcome that can help build and maintain a strategic, resilient team: defining when the work is done.

Most of us work in complex organizations that have multiple dependencies that can make it hard to know what the work is and when is it truly done. For example, many teams and management are far removed from their customer, which makes it difficult for the team to know when the customer has authentically used what was made, and if they liked it. Yes, there are fancier words to represent customer satisfaction; let's use the word *like* to keep it simple. How can the team or its management know their work is done if they don't know that their customer is using it and likes it? Also, another part of the organization that didn't know about a new product or feature the customer is using might be negatively impacted by it. Now all teams and management are scrambling to communicate and adjust when it all could have been prevented up front. Sometimes, the entire organization is compromised, and the amount of scrambling is exponentially large. Strategic, resilient teams work with management and any stakeholders to define when their work is done. It's a heuristic process, refined over time.

With customer value in mind, the team and management collaborate on what it means for their work to be done. Since this group already knows what coaching means at their organization and who their customers are, they can define the key steps in what it means to be done with a work item. Many work items are unique; the definition of done will be, too. And, expect them to change over time. As the team continuously learns and improves their work, their definition of done will reflect what has been learned.

Sample Definition of Done

An innovation coaching team has a work item called *client discovery*. This work item requires a coach to contact the person requesting innovation coaching, and to schedule time with them to understand what they need, create the output from that meeting, and then share a coaching recommendation with the client.

The team determined the following definition of done:

- Add work to team intake system.
- After work has been pulled from the team board, meet with requester to perform discovery questions using team discovery agreement.
- Create discovery output and coaching recommendation on team template.
- Load discovery output and recommendation to team portal.
- Present discovery to requestor.
- Discuss next steps, including prioritization and order of work.
- Update the team board.

Getting all these steps done informs the team and management of key information about this potential work in a consistent way that everyone understands. Having the all the details of the client discovery enables the team and management to better understand, prioritize, and order their work so that they can deliver relevant value to their customers. Without these details, the team is merely speculating about their clients' needs.

Sometimes clients are in a hurry. In their haste, they will contact the employee-coach team management to request an expedited discovery. For example, the main stakeholder who has all the answers for the client discovery is on vacation for two weeks. The team is asked to just get started now and do the discovery part later. The need to have action, to show activity for both the client and the coaching team is strong. Yet, skipping steps in the team discovery puts the coaching team at risk of missing what matters most to the customer. Skipping strategy for the sake of action rarely has a happy ending for either side of a coaching relationship.

> Skipping strategy for the sake of action rarely has a happy ending for either side of a coaching relationship.

Collaboration is the follow through of a strategic intake system and in managing a backlog of work. Successful employee-coach teams meet with their management at least once a week to review their backlog of work. As a team that continuously learns, they use any new data or information as a lens to review the priorities and order. At the beginning, this work may be time consuming; it will improve with experience. Having an informed view of the work in its most current state informs decisions and helps create the environment for team resiliency and customer value.

Build a Customer Feedback Loop

A feedback loop is a way, or several ways, to improve and measure customer satisfaction (value!). In a feedback loop, the employee-coach team and the client are in a constant state of sensing and adapting to what is needed, while staying grounded in vision and strategy.

What does all this mean for an employee-coach team? They are operating their team with strategy, continually sensing and adapting to what is needed to deliver customer value. Employee-coach teams with a well-maintained feedback loop have a better chance of being strategic and resilient.

The strategic work the team and management did together in forming the team, or refreshing the coaching team, will help inform and provide focus to the kind of feedback needed. In the strategic formation, the team and management established *"What does it mean to be successful?"* and, *"How will we measure success?"* Building on that work, the team and management shape their feedback loop with the answers to these questions:

- What will we measure?
- Why measure these things (connect to vision and strategy)?
- How will we measure?
- How often will we measure?
- Who does what?
- How will we learn from the data?
- How will we keep our work visible to the team, management, stakeholders, and clients?

Give this work time and keep an open mind about creating and maintaining your feedback loop. Your loop likely has different parts to it, both formal and informal. While satisfaction or other types of surveys are not a bad idea for a customer feedback loop, they shouldn't be the *only* idea. There are many approaches and tools available to you.

A strategic team who is continuously improving themselves can usually find creative ways to measure customer value. They need the time and space to create and test their ideas. Innovation rarely works when it's shoved into the scraps of the team's time. Make the time. Invest in follow through. Use these considerations to shape the guidance you'll give your team and management on using and improving the feedback loop:

> A strategic team who is continuously improving themselves can usually find creative ways to measure customer value.

Feedback Loop Guidance

1. Have a solid definition and understanding of the employee-coach role in your organization. A team with an identity crisis will struggle to be strategic and resilient.

2. The expectations of your customers won't be aligned to this definition, unless you prepare them before offering coaching services.

3. Approach customer concerns with a curious mindset.
   Seek first to understand what the struggles and victories are all about. Approach the problem or opportunity as if hearing it for the first time and avoid glossing over the details to arrive at a recommendation.

4. Building on curiosity, work to meet the customer where they are. Your customer is not likely to be as advanced in the transformation or change happening in the organization as the coach team and management are. Their current state or current place on their journey has a large impact on their perspective of coaching and the value of it. Gain understanding on where they are, so that you can develop some perspective and empathy.

5. Be wary of over-the-top, positive, glowing responses.
   Change and transformation often creates artificial buy-in, for all kinds of reasons; many driven by fear. Dig into the glowing data as much as you investigate the negative data.

6. Understand the team's vulnerability.
   Negative experiences do not automatically mean the employee-

coach did something wrong or behaved inappropriately. Employee-coaches are extremely vulnerable to ancient corporate history that they may or may not have been associated with, as much as they are to their role being misunderstood. Be curious. Get the facts.

7. Surprises in the data should be few.
   An employee-coach team that intentionally collaborates, makes their work visible, inspects and adapts their work often, and prepared their client well should rarely experience surprises in their feedback. The team and management know their strengths and weaknesses, and client challenges and opportunities. All's well that begins well.

8. Avoid secret feedback loops.
   Many times, management is seen as a secondary or even covert feedback path, which creates miscommunication and a negative environment for all. Existing organization behaviors are a strong current against the feedback loop; management feel this pressure the most. Resist the temptation to fall into former habits.

9. Refine your feedback loop over time.
   Make it part of other team strategic discussions.

The bottom line is, build a feedback loop. Most organizations know what they need to do; they can't get the agreement, momentum, or courage to do it.

Build Team Continuous Learning & Improvement

For a team to be strategic and resilient, they must have an environment that offers time, space, and safety to continuously learn and improve themselves. Often, the organization vision supports continuous learning and improvement but the complex environment and pressures to deliver work make it difficult to uphold them.

In the team strategy session, it was already determined how the team wants to grow and why. Now it's time to walk the talk of making continuous learning and improvement a reality, becoming a mindset and a way of working. Expect a lot of guidance here, because continuous learning and improvement is the essence of the employee-coach team.

Actions and Mindsets that Support Continuous Learning & Improvement

Strategic Investment

Remember, we are focused in this chapter on the employee-coach team operating with strategy. Your team aims to be an asset to the organization, and activities to improve the team are investments in them. Make any requests for learning and development funds or time away to learn, a request for *an investment in the team.* Even if your organization doesn't see teams this way. Many organizations are well-intended with their vision and values, but few actually live out the details behind the words. Don't let this current state hold your team back from working to invest in yourselves. Look around your organization for a team or an area that is doing what you aim to do. If you find them, talk with them about how they invest in themselves. If there are no teams like this, be the first. Use the team vision and strategy, connect it to the overall company strategy, and sell your management on the impact of the investment. Be ready to repeat yourself, a lot.

> Many organizations are well-intended with their vision and values, but few actually live out the details behind the words. Don't let this current state hold your team back from working to invest in yourselves.

Gather Often

Find ways to intentionally develop and learn as a team and as an individual. This means learning needs to be the star of the show, rather than getting the scraps of a meeting that had a different purpose. In other words, when the team meeting has an extra 15 minutes, don't retrofit a learning item from the team's backlog.

Create and experiment with team gatherings dedicated to continuous learning. Also, when you do have a learning event, resist the urge to have this time swallowed up by a delivery hot topic. Expect this effort to be challenging and imperfect. Keep trying experiments on what works for the team.

Cultivate a Continuous Learning and Improvement Mindset

Mindsets don't change or develop overnight. Through strategic investment, regular practice, and manager support, your team will begin to experience continuous learning and improvement. Repeated experiences that are positive and developmental help grow or strengthen a mindset. Once a team and manager feel the impacts of this work, it's likely they won't want to work any other way. Every team and strategy is unique; consider these options:

- Team creates ways to keep learning in front, such as a one liner that is used in daily interactions, such as, "What question should you be asking right now?" Or, "And what did you learn?"
- Make it safe and encouraging for people to give and receive feedback. Try using the method *glow* and *grow*. *Glow* is the strength and the value; *grow* is the weakness and the growth opportunity.
- Integrate the learning strategy into the start of team reflections, and other key team collaboration.
- Make learning goals and successes visible in the team's workspace, physical or virtual.
- Keep and encourage a curious mindset for the team and management, rather than all-or-nothing-thinking. Just because one

approach didn't work doesn't mean nothing else will work. Some of the most defeated moments as a team are when a manager or stakeholder sees one moment that didn't work out as failure, and then the entire program is deemed a failure.

- Encourage continuous learning into employee-coach individual goals. Management can do this by designing their own goals with continuous learning and improvement, and then sharing them with the team.

These recommendations are not new; what's new *is actually doing them and the follow through.* Many teams and management make it past *trying*, but then a disruption comes along and knocks them off course. Then, they are not resilient enough to get back on track; it feels impossible and time-consuming, and then all work is abandoned. A strategic, resilient team that operates with strategy makes their continuous learning and improvement bulletproof. Disruptions are handled by the team and management, and then it's back to the business of learning as quickly as possible.

Healthy Patterns for Learning

- Avoid favoring specific people for the team's learning investment. So many times, the experienced coaches and/or the ones who are not afraid to ask, are the ones who are favored for learning funds. Nothing steals from a team more than when favorites are consistently given opportunities that others on the team are not.
    - Keep the concept of a cross-functional team a priority. The team and management must work together to ensure pairing, shadowing, and other cross-functional learning happens for all team members.
- New team members are equal team members.
  Often, new team members are either sent alone into the fire without any support, or they are sent to the corner to read documents and

books. Start growing your new team members' continuous learning and improvement mindset on their very first day.

- Always make time to share back learning with the team. Whether it's the team or management, a continuous learning mindset means there is *continuous sharing* of learning. Strategic and resilient teams make regular time to share experiences, learning from formal events, and field visits. Many teams make the activity of the share back part of the definition of done for attending a learning event. Whatever makes sense, make learning part of being on the team.

- Management of the coaching team: One specific way management can help is not to accept the default option handed you by the organization.

  Your team is different and has a great deal of responsibility and potential to impact many people. Push for innovation rather than the usual training budget mindset. Sometimes it makes strategic sense to contact your partners, such as learning and development or human resources, sometimes it doesn't.

Growth, Competencies, and Team Maturity

All's well that begins well, and, everything can be improved. This section is specifically focused on the team's view of individual career development. When the team and management had strategic events, they discussed the topic of team growth and learning. Specifically, the question was, *"How will we continuously learn and improve together and from each other?"*

The employee-coach team may not have a formal career path inside the organization, but it most definitely should have a *growth path*. Let's unpack this thought for a moment. Employees of all backgrounds often desire or even expect a clear career path for their job: a clear path of growth that is often reflected in responsibilities, competencies, and levels. The level-fixated environments that many organizations have created over the years are a big reason for these expectations. Levels are not always bad, but many times they create the environment for poor management behaviors and anti-growth behaviors. The topic of career paths and levels is rich. For the scope of this book, here's the bottom line: A career path and career development are two different things.

> A career path and career development are two different things.

Focus first on growth and career development. Keep it lightweight and very intentional, for the employee-coach team only. Even a lightweight, intentional focus can go off the rails, becoming an over-processed, detailed list of required tasks and achievements required of all. Use the outcomes of the team strategy event to guide the structure, focus, and content of the team's career development. Consider the following ideas for discussion:

Employee-Coach Growth Topics

- Specific coaching competencies
- Specific coaching experiences and activities
- Thought leader within the organization
- Thought leader beyond or outside the organization
- Soft skills
- Relationship skills
- Industry specific skills
- Customer focus
- Leader skills (remember, this is different than management)
- Teaming skills
- What part of growth does the employee own?
- What is expected of management for employee growth?
- What is expected of the organization to support growth?
- If you must make a metric of coaching competencies and skills, make it strategic!

Some employee-coach teams make a coaching maturity model out of these considerations, dividing them by beginner, intermediate, and advanced (or establishing, intermediate, and expert). Be cautious about calling one group a beginner or another an expert, or any other buzz word that will encourage a client to only want to work with one type of coach. Levels, words, and published hierarchies only steal from the team's credibility and create separations within the team that usually turn negative. True coach maturity models are about experience, knowledge, and outcomes achieved.

This section is not trying to diminish the superpowers of the experienced coaches over the inexperienced ones, rather, to promote cross-functional and continuous learning. Nothing will ever be completely even, but there is value in having intentional, focused growth that everyone on the team can see and access. When only a few team members have all the knowledge, the team is far less strategic and resilient in difficult times or when there is a great opportunity. Another way to say it: a team is only as strong as its weakest link.

Finally, no matter how your team organizes its growth, the best approach is to keep the details with the individual and the team, not broadcasted to clients. For example, saying, *"We're giving you Bill, our beginner coach, to work with you"* might sound like Bill is the last choice. What the client doesn't know is that Bill has deep skills in what the client does. So, even though he is a beginner coach, he understands the team's work very well.

Adjacent Disciplines

One of the most critical career development activities all employee-coaches and management should practice is *adjacent disciplines.* From Microsoft thought leaders to The Karate Kid movie, people who practice strengthening their adjacent disciplines have improved their careers in ways they couldn't have imagined at the start.

> One of the most critical career development activities all employee-coaches and management should practice is *adjacent disciplines.*

Adjacent disciplines are those which grow skills, experiences and knowledge in areas that *surround employee-coaching.* When all the individuals (including the management) on the team practice adjacent disciplines, they create a collective force of team knowledge, skills and engagement. Over time, a team and its management become stronger in problem solving, innovation, and the overall flow of coaching work through the team.

Studying adjacent disciplines is a personal journey; encourage team members and management to share what they are comfortable with as part of the continuous learning focus.

Employee-Coach Adjacent Discipline Examples

- Facilitation: being a great facilitator makes space for coaching.
- Virtual technology tools: expertise using the latest tools makes coaching relevant.
- Public speaking: confidently speaking to large groups enhances credibility.
- Drawing: artfully expressing concepts makes a coach meaningful and memorable.

Build Credibility

One of the single greatest challenges for employee-coaches and their team is establishing and maintaining credibility. A credibility problem is a trust problem. And, it's not limited to employee-coach teams. Perhaps this is why there are so many books, seminars, and blogs about the topic of trust. Any team and managers of a team planning to stick together for the long run would benefit from studying and discussing their credibility and trust. With so many experts and published works, the difficulty will be in narrowing down the options that are most interesting to the team.

The root of many of the employee-coach team's credibility challenges are related to the organization's environment and culture; not the coach lacking ability, and not because the teams or management are bad. The environment of an organization consists of how work is funded, how people are compensated or rewarded, how management lead (or not), how work is prioritized and ordered, levels of careers and management, just to get you thinking. One of the main outcomes of an organization's environment is its culture. An organization's environment and culture often cast a shadow on an employee-coach team's credibility and then of course, its individuals. This is where having a strategy and strong sponsorship of the team makes all the difference.

The organization's environment is not the only challenge to a team's credibility and trust. The team and management often get in their own way. Let's explore common reasons why employee-coach teams don't have the trust needed to be effective.

Misunderstanding of the depth of the change or transformation at hand.

- We already understand what you say you are here to coach us on.

Poor positioning of the change or transformation.

- Our work is different; we don't have to change.
- That coach was ineffective; they were always encouraging us to find ways to change things, when everything we do is already as optimal as it can be.

Poor positioning of the employee-coach as support for the transformation.

- No one told us we were going to have a coach.
- We don't need someone to tell us how to do our jobs.
- We are all experienced in the methodology and mindset. We don't need a coach anymore.

Misunderstanding of the employee-coach role.

- Aren't you just a project manager?!
- Can you schedule and facilitate our meetings for us?
- We thought having a coach meant we got a glorified (project manager/business analyst/engineer).
- Can you keep an eye on Fred for me and let me know how he is doing?

No sponsorship

- I am a busy manager and I don't have time to meet with you about…anything.
- Our management said we don't have to listen to you or do the changes.

Low sponsorship

- We heard you were coming to our team, so we got everything done ahead of time. We don't need help.
- Sorry, it's taken so long for me to respond about coaching our team. We were really busy and I have a short vacation coming up; can you schedule something for us to talk late next week?

Bad coaching experience in the past means all are bad.

- We had coaching from your team in the past. It wasn't valuable.
- We had a coach from a consulting firm, and it was more harm than good.

Organization is inexperienced with being coached.

- What do you mean, *"What do you want to learn?"* Just tell us so we can get it done.
- That coach was ineffective; all they did was ask us questions when they should be telling us what to do.
- That coach was too much of a purist. They didn't understand that our problems are different.

Legacy relationship experiences that were not positive.

- There is no way I'm being coached by Brett. We were both on Project T together and it was a total failure.
- I know that person is a level below us. They will never understand our work.

Jealousy

- I applied for the employee-coach role and was not selected. I was told I was inexperienced. I'll show them!
- Great, only the shiny, new teams get coaching. Meanwhile we are ignored.

The following cause of low trust occurs within the employee-coach team relationship system. The impact of this dysfunction is felt within the coach team and can indirectly and negatively impact the client relationship.

A Room full of Experts

That meeting is so depressing.
We can't self-organize out of a paper bag.
We talk and talk and nothing ever gets decided!
I know I can't skip the meeting, so this week I coped by shopping for boots.
We are a great group of coaches; why can't we make any decisions?!

An otherwise healthy, cross-functional team with growing knowledge is a fantastic thing. Yet because the coach team consists of experts, there is a high probability that they will get in their own way. This is a very common finding in all types of teams of experts, and often affects management teams, too. The impact of a team unable to get past themselves can be deflating at best, and renders the team unproductive. Even with a working agreement and good relationships, the Room Full of Experts impediment troubles many teams.

Even with the most focused and strategic beginning of a meeting, a room full of well-intended experts are often challenged to remain focused. There are all sorts of environmental and interpersonal reasons for an expert to lose their focus. The experts spin their wheels in minutia, agree on things until it's time to make a decision, and then throw on the brakes. Let's look at an example:

Example - Virtual Call full of Experts

During a weekly team meeting held online, one of the topics is an update on the refresh of the team's intake system. Last week, the team worked together to identify some improvements, and agreed on which ones they would carry forward. The team agreed to have a sub-team, work on these improvements, and then report back to the team.

Now, the sub group is reporting back to the entire team. As Harpreet, a sub-team member primes the discussion by sharing the context of the last meeting, she is interrupted by Alex, a team member.

"Wait, didn't we talk about this and settle it the last time we met? Why are we talking about it again?"

"I don't think that was quite it," another team member chimes in.

Harpreet then feels compelled to explain the context of the work from the beginning, again. While this happens, those in the meeting who knew this was an unnecessary review say nothing. Instead, they open their phone to check the weather for tomorrow.

"Thanks for explaining that, Harpreet. I'm not sure these are the right improvements," Alex says.

With the support of other sub team members speaking up to try and stay on topic, Harpreet presses forward. She anticipated a lot of questions about what the sub-team created, but now it feels like they aren't even looking at the work. Was it something she said?

"I have an idea," Alex declares, and then begins to share it.

> Other team members forget the goal of the subgroup, in fact forget all about the subgroup, and engage in Alex's idea. Some team members like it, some do not. New ideas form that spiral away from the original need.
>
> Harpreet and the sub-team aren't sure if they should shut down the entire conversation, let it go for a while and then reel it in, or just let it go all together. They wonder, is this the same team as the one that charged this sub-ream with the work? They included everything the team asked for; why is it so disconnected?

In this example, the team members are not evil; they are caught up in their expertise and the need for it to be visible and are unaware that it is derailing the team. Experts are not always arrogant, but many do find it rather irresistible to talk about related experiences and knowledge. Especially when they are together in a group, where expert energy feeds more expert energy.

When the room full of experts effect routinely emerges in the team meeting; the team struggles. The way the team goes about getting through their agenda is ineffective and has emotional fallout for some or all of the members. Then, they leave the meeting wondering why they feel exhausted and unproductive.

There is hope! Fixing this team impediment is not easy, but there are some tactics a team can use to reduce it. If you have a coach of coaches, and they are not already involved, enlist their help. Use the following guidelines to reduce or remove this behavior from your team.

Awareness

The team must be aware that they get in their own way if they are going to improve themselves. Awareness is usually built over time, not a switch that is suddenly turned on full power. No matter how the team decides to raise awareness of the problem, it's really important to not label it as bad or good.

It's not about blaming the person or the team; it's about raising awareness of something that is part of the team's current state.

Call it Out

Awareness of any team impediment will only get the team so far; when it happens, team members must speak up. That action can be easier said than done. Growing a safe environment for all team members to openly push for change and innovation is done over time, one meeting at a time. It's important for the entire team to help create this environment, rather than the same one or two team members always speaking up. Everyone needs to feel safe to call out an observation and request for change, even if they are not willing to do it. If you feel this is a team impediment, consider making *call it out* part of your work or team agreement.

Celebrate Progress

Seek progress over completely eliminating the problem. When you do catch yourselves as a team, or someone speaks up about acting like a Room full of Experts, celebrate it. Raising awareness and self-correcting are both progress toward the long-term reduction or removal of this impediment. And, progress and growth as a team.

Sponsor and Lead Change

You know by now that successful employee-coach teams need strong vision, strategy, and sponsorship. The coaching team's vision and strategy is not just to ground the team and management in purpose and intent. It's also for outward-facing messages on positioning the coach team in the change or transformation, and, communicating the value of employee-coaching. When the employee-coach team's management and stakeholders relentlessly communicate in interactions throughout the organization, the steady drumbeat of the value of the coach team is eventually heard. But it doesn't stop there.

The people and team(s) leading the change that the coach team is supporting also must be well-versed in the coaching team's vision and strategy. No one is expected to walk around reciting the vision and strategy verbatim; instead, to deeply understand the connection. Change leaders know how to internalize the vision and strategy and then be able to effectively communicate the change in a compelling and inspiring way.

While the change team(s) were hopefully identified as stakeholders in the team's and leader's strategic events, it's worth calling it out again now. These change leaders have the advantage of *position* in the environment mentioned above. They usually have the attention of senior executives and management, from both the delivery of the change at hand, and the enablement of the change. They are usually the best people to champion the employee-coach team as strategic to the change, now and an in the future. If the employee-coach team can be seen as strategic, they will have better chances of being seen as credible and ultimately, trusted.

If the change or transformation team(s) and management have their own credibility issues, there is opportunity and hope to improve the current state. The employee-coach and change teams and management can join forces to gain critical mass that cannot usually be ignored by senior management. This kind of effort usually requires the teams and management to slow down to plan and create a strategy that will ultimately help them gain speed. Change and transformation is relentless work that requires a sense-and-adapt mindset. If after several attempts and experiments the change and coaching teams cannot influence senior management, the need for a different conversation has arisen. It's time to talk with those who originally invested in the change and transformation work, to determine if the effort is still viable. Many teams in this situation (yes, unfortunately it's very common) evaluate the situation with rigor, such as A3 thinking. It's not necessary to do a full-blown problem analysis, but supporting data will help your cause. It may be enough to gather the data and create a bottom-line statement for executives such as, *"Our organization's actions suggest we don't need to move faster right now."*

Encouragement for you, and the bottom line, one more time: An employee-coach team that is not properly positioned as a strategic, impactful part of the desired change is not going to be trusted.

Prepare Clients to be Coached

Prepare Clients

Even if you are not the client's first coach, set the stage for a strategic and meaningful relationship. Offer a preparation document and discussion as part of the team's definition of done for a discovery of work. Make time to help clients really understand what coaching means in your organization, and what they can expect. Begin with intention and goals of coaching to get the most value out of the engagement. Adjust as you learn and grow together.

Show the client how the team's work and yours connects to the vision and strategy of the change or transformation work, and, ultimately the enterprise. Use storytelling and your team's marketing artifacts. When the employee-coach team consistently uses the same messages with new clients, they gain credibility, or at least plant the seed for it.

Establish Clear Outcomes

Work with your client to define what they want from coaching. Contracts, discovery agreements, and statements of work are likely necessary parts of the system you and your client are in. Beyond the documentation, have an actual discussion.

Ask the client questions like these:

- What are your expectations of me as your coach?
- What outcomes do you want to achieve?
- How will we work together?
- What defines success for us?
- How will we measure progress?

Stress to the client that quality individual and team coaching experiences take time. Clients who give their time and focus to coaching are far more likely to have successful experiences. Slowing down to talk about what was tried before, the current and the desired, future state is critical to shaping outcomes.

Getting and establishing buy-in from management is important to introducing and positioning an employee-coach team. Also important, the client must make space for learning in daily work. Encourage the client to communicate to stakeholders that they are committed to creating sustainable change and the work demands consistent coaching. This prepares them for the reality that their work priorities and coaching journey will at first collide and then integrate. Over time, engaged management and teams will seek out opportunities to improve their work and processes through coaching.

Keep all the ideas organized and visible for you and the client. There are many options and software apps available to make the work visible and adaptable to change. Having a view of the work will make it easier to prioritize and order the work; knowing what the most important thing is to work on also reveals *what is not important.*

The backlog of coaching work is really the easy part. The challenge is in getting the client to see the value of slowing down to talk about the work before doing it. Most managers are very supportive of coaching, until it affects their favorite program.

Be Honest

Establish the confidentiality needed to effectively work together. Encourage your client to be honest with all they can, so that you have as much context as possible to help them. A common statement that helps establish the need for honesty is, *"The more context I have, the more strategic my coaching will be."*

Honest context of the work and the team's current state often reveals problems and opportunities that have been avoided in the past. A true coaching effort should help the client move past or through these challenges once and for all. The best part of this work happens when the obstacles are gone, there is new space for the team to innovate and add value to the organization.

Own it!

The employee-coach service is founded on ownership of learning by the client. When a person owns their learning, they are willing to learn out of their own desire to understand. Make it clear to the client that ownership of learning is a mindset. Offer empathy and encouragement to them, because a learning mindset can be difficult to practice when there is a lot of work to do. Over time, with your encouragement, change can happen. You want the client to really think about what you are asking of them, so they can apply it to help them create their next, right move. Encourage and expect small steps and experiments at the beginning of this journey.

Successful employee-coaches do not prescribe how to do work or give solutions to problems; they challenge with questions, share their insight, and offer advice. It's up to the client to apply what they hear from a coach to their knowledge and experiences.

When a manager and team own their own learning, they begin to make a strong contribution to the change or transformation, ultimately impacting the entire organization. They become a force multiplier for change.

Experiment

Avoid all-or-nothing thinking within your coaching team and with your clients. Encourage a small-scale experiment with your client, so that they can see the value of coaching in a small effort that has perceived lower risk. Co-create experiments with your client, which gives them ownership and helps build trust.

Transformation and change work is undulating. With uncertainty as the norm, having an enterprising, experimental mindset with coaching offers a chance to create a cycle of taking small steps, inspecting and adapting results. This approach helps create a coaching experience that is meaningful and sustained. Encourage your client to use this inspect-and-adapt cycle.

If the organization is new to experiments, or the word makes them uncomfortable, you will need to build more trust with them to encourage participation. Help the client see that the small-scale experiment won't likely cause any big problems.

Encourage Perspective

Encourage your client to see the value in coaches with different styles, perspectives, and advice. Most coaches won't run a client off the road. Most. Remember the quote, *"Ask a different doctor, get a different opinion."* Well, most doctors have the intent to help a person be as healthy as possible; the same applies to coaches. When the intent of coaching is consistent, there is great value in the client experiencing different coaches and their insight.

Having perspective comes back to ownership. The main ingredient for a positive experience when there are multiple coaches is for the client to own their learning. Ownership makes it safe for the client to look at the different coach insights, and then apply them to their own perspective and knowledge.

Stay Connected

Educate and encourage the client that a successful coaching experience needs the time and space for connection to grow. In the beginning, it's usually the coach actively seeking the connection, and modeling the desired outcomes.

It's up to the client to respond and participate. Make a mutual commitment to connect, with the intent to get the value out of coaching the client wants and needs. The client and coach commit to each other that they will work to connect and share expectations. Grow this commitment over time, with consistent work and authentic discussions.

How does this work? Build a cadence of connection from the beginning, and then sense and adapt this connection over time. Time with a coach can't be delegated to others. Most coaching takes more than a few meetings to learn new capabilities or unlearn behavior patterns or ways of working that are no longer serving the team or the organization.

How to Approach a New Client

So you're an employee-coach of a team with a vision and strategy, a solid intake system, and a definition of done. You've been assigned an area of the organization or a client to work with. Now what? All's well that begins well. Let's look at effective, strategic ways to approach a new client.

Favorable Introductions

The best way for an employee-coach to begin well with a client is with a favorable introduction to the management (as always, includes stakeholders) and team. A strategic and resilient employee-coach team knows how to give their people and their clients the best possible chance at a healthy relationship. It can be tempting for the client to skip this step, especially if there is urgency to get started. The truth is, the introduction *is* getting started.

A strategic, consistent message used by the employee-coach team management helps the team position itself in the organization. Some situations may need adjustments, as long as the message doesn't go against the employee-coach team's vision and strategy.

These questions will help prime your team's discussion to create a strategic message:

Potential Introduction Topics	Strategy Supported by Topic
Why are there coaches in this organization?	Connect the overall vision and strategy of the organization to the role.
Why do *we* have a coach right now?	Connect coach team and change strategy to client.
What is a coach for us?	Connect coach role definition to client.
How do we get value from coaching?	Prepare client by sharing expectations.

Other guidance for the message:

- Short and to the point.
- Little or no corporate speak.
- State at a high level what the coach will do.

Example 1

"Hello, I am Ava, and I am a coach on the Macadamia team. As you may know, the XYZ change is moving us from project to product thinking, and intends to blend discovery, delivery, and learning. The coach team supports this change by working with teams and management on what they need to change the way they work. I am dedicated to your team one hundred percent for three months, to coach and consult product thinking for the team and individuals. Let's talk about getting started."

Example 2

"This is Olivier, a coach from the product coaching team. We all know our company is transforming our product delivery strategy, and the product coaching team was created to support that work. Olivier will coach our team and us as individuals, so that we can fully support the transformation and be a part of it. Olivier is one hundred percent dedicated to our team for the next two months, and we intend to make the most of that time. Expect to see a lot of Olivier in your daily work and in our meetings, so we can learn from him."

Unfavorable (!) Introductions: How NOT to do it

- "This is Andrew. You can ignore him as he is here to just observe what we do and suggest things, but I didn't want everyone being confused about who he is."

- "I would like to introduce you to our new coach, Priya. She is here to deliver change and make us better.

- "This is Jeff, our new coach. He is here to validate that we have all been doing it wrong and that our current way of working is simply not good. He will fix us in no time."

- Client Manager: "Team, meet Brendan He's here to make sure we are doing design thinking right. He's got a long history of coaching teams like us, so I think he will fit right in."

- Coach Brendan: "Thank you, manager. I'm glad to be here. I look forward to helping your team.
  Client Manager: "We look forward to having you around. Now, onto other business..."

- Employee-coach manager: "Client, I've assigned Vijay to your area as a process improvement coach. He has a lot of experience with what you are working on."

Generously Curious

Curious coaches are open minded to what's happening around them, no matter the level of experience they have. They know client behavior patterns of their trade well, and yet they hold back on deciding...anything. Free from attachment to past experiences, emotions, or specific outcomes, a coach can be generous with their curiosity to honor the client's uniqueness.

A generously curious coach knows how to share the intent of their work with the client at the beginning, so that there is shared understanding of this coaching pattern. They are able to show the client that seeking to understand the current state is in fact getting started on coaching and refines the aim and intent. This coach also knows how to resist the pressure from their own organization to hold back or even skip the step of getting to know the client.

Generously curious coaches embrace the strategy to *meet people where they are*, and then show the client where they can go. Meeting people where they are means asking questions, listening intently, and then being unassuming about the client's current state and what they have experienced in the past. The curious coach also knows that it takes time to get a deep understanding of a person or a team. They know that no matter how it looks, people are doing their best with the information they have. They generously invest time in the client relationship, with the future in mind.

After getting a deep understanding of their client's current state, the coach can then show the future to the client. Showing the future isn't a mystical thing, rather an exercise using the context of what the coach discovered to show the client what is possible. *You could be here.*

Neutral

Neutral coaches understand that freedom from attachment gives them space to be their best and deliver client value. They are comfortable in ambiguity and actually expect it.

Change is in the nuance. A neutral state of mind finds the nuance. When all the obvious problems and opportunities have been tackled, the critical work that remains is much harder to see. Without attachment to the drama of the last team meeting, a neutral coach can see so much more than others can. This coach can see the systems-view-big-picture, and *nuance*: the little moments and micro-decisions that over time, add up to impactful change. Successfully leading and coaching change happens in these moments, not usually in the big reveals of training and educational events. The neutral coach will not only *observe* nuance and use it in coaching but also seek out and *create* nuance by using empathy and neutrality to open the door to a conversation.

The neutral coach aims to be respected rather than well-liked and knows that their organization may not be ready for this coaching pattern. They also know their neutrality may be unpopular some days and treasured on others. Their neutrality helps them to be unflappable in high drama moments. A tense moment can be neutralized with the right question or statement, often resetting the mood or conversation.

The neutral coach is not manipulative; there are no secrets in a neutral position. The neutral coach makes it safe for clients by stating the path and the intent of the coaching work. *They say what they will do, and then they do it.* This pattern doesn't mean revealing all your notes and tactics, but enough to clearly and easily show what you intend to do, what is unknown, and what is out of scope.

Encouraging

Effective coaches can authentically encourage their clients to take a step forward from where they are. They know how overwhelming change can be, so they have empathy and encourage their client with small steps that add up to change over time.

When things start to spin, or go into difficult territory, the coach encourages the team to remember what they are doing and why. When helping the client get unstuck, the encouraging coach is doing it in a way that teaches them how to help themselves in the future.

Relentlessly Strategic

Coaches who relentlessly connect the client's work back to the higher cause of their team and the organization have sustainable change and the future in mind. The strategic coach helps clients to examine problems down to the root, not just at the superficial level.

Also, the coach and client apply systems thinking, which is a perspective that includes all systems related to the work, rather than a single part. When clients apply systems thinking to their work, they are better able to see patterns and events over just isolated data points of interest.

The strategic coach is also always looking for opportunities to subtly market the value of coaching and success of the client. The client is often too deep in their own change process or transformation to be able to see the progress they achieve. When the strategic coach is thinking about the higher purpose, they are better able to help the client see growth happening and are able to share the success with the greater organization.

The strategic coach is also always looking for learning opportunities for the client and the coach team. They are hungry to continuously learn and improve themselves and the coaching team.

Coach Failure Patterns

This section reveals the *what* of coach failure patterns. Reasons why coaches fall into these and other failure patterns are varied from personal to organization environment. It takes reflection, and likely a coach of your own to work through the reasons why a pattern exists, and then have a plan to change it.

Victim

Of all the reasons, rationale, and justifications for coaching in a failure pattern, victim thinking may be the most destructive to both the coach and the client. Thinking and saying, "*I had no choice.*" is not professional coaching. Maybe there is no simple or easy choice, but there is always a choice. The work of being a coach is often about first looking inward, and then encouraging a client to push past the temporary discomfort to create a better future. A mindful coach knows how to reflect, stay unattached, and take some of their own advice.

Domestique

The Domestique coach does everything for the client, in hopes they will own it later. There is value in a coach doing a small amount of the client's work to show the client how to do it, and to give them momentum. This approach must be done with intent and clear communication, so that the client understands you as the coach are only showing them how to begin. They must own the rest (own their learning), and you will be there to continue to coach them.

The Domestique schedules and facilitates meetings on behalf of the client. Rarely should a coach of any mindset, method, or framework be leading an entire meeting. The client must own what you are helping them to learn, collaborate, and decide.

A Domestique takes on communication tasks that should be owned by management and stakeholders. They say yes to everything.

Judge

The dissenters and resistors of the change or transformation are bad.
The client is bad because they didn't do what you recommended.
The client is good because they laugh at your jokes.
The client is good because they do everything you recommend.
The client asks too many questions, so they must not be smart.
The judge is a coach who is uncurious about...everyone and everything.

One of the Team

The will to be liked is stronger than the will to be a respected coach.
No longer a curious and neutral third party, the employee-coach is one of the team.
Misses nuance of change because s/he is thinking like a team member rather than a coach.
The coach undulates with the team's emotions, victories and losses.
The coach is so embedded in the team that they can no longer see as much of the big picture as they once did. Pickled!

Know When to Say When

Now that you've seen the most common coach failure patterns, its time to talk about necessary endings with your client(s).

You can't win them all. Regarding credibility and trust, an employee-coach definitely can't win them all. This doesn't mean anyone is bad. Remember, a lot of problems in the organization's environment can really hold back a change or transformation, and the people who support it. Sometimes even the best sponsorship, support, and positioning of an employee-coach team are no match for a strong environment pulling attention, actions, and work in another direction.

Step 1: Assess and Reflect the Entire Situation

This means reflect on *your behaviors and actions,* too. Strategy is still alive in a *know when to say when* situation. Some teams use A3 thinking or a simple, lightweight hypothesis to help better define the problem. These methods are helpful but can sometimes be distracting and slow down what could be a simple decision process.

Whatever strategy your coach team chooses to solve problems like credibility and trust, it's important to first reflect on your own actions and intentions. With a *learning* lens, not a *failure* lens, make time to look within to examine what you did well, and what you could have done better. You can also get help walking through your experience from a trusted colleague or your mentor. No mentor program within your coach team? It's worth a discussion with your team and management at your next retrospective.

Step 2: Communicate the Problem to the Employee-coach Team and Management

Since you are a strategic and collaborative team, the problem is not like a surprise to anyone. Share the situation and your recommendation with the team with your coach of coaches present.

Step 3: Collaborate with your Team and Management to Determine Action

Collaboration gives the entire team a chance to learn from the problem, and provides a wide perspective on the problem. While it can be uncomfortable to get the team involved in an already uncomfortable situation, keep a systems thinking perspective. This single problem rocking your world is more connected to the rest of the coach team and their work than you might imagine.

Solve the Problem

Use this strategy when there is a specific barrier that if removed can solve the problem.

- Possible ways to try to solve the problem:
    - Use can-if thinking to help arrive at a lightweight hypothesis. *"I can build trust with this client if I get more time with them."*
    - Try a small experiment to test the lightweight hypothesis. *"I got commitment to join the team's planning meetings and their management's planning meetings."*

Cope with the Problem

Use this strategy when there is no option to leave the client. Explore ways to cope with little trust.

- Possible ways to cope with the problem: assign a different coach to the area. This solution is not in the *Solve the Problem* section because there is very likely an organization or management reason the employee-coach wasn't successful. Switching coaches may be very effective, and, it is coping, rather than solving the problem.

Collaborate Through the Problem

Use this strategy when the client is aware of the trust problem(s), and is willing to work on them with the coach.

- Possible ways to lead through the problem: Spend time to collaborate with the client, including their management. Seek to understand what is holding back trust. Be curious.

Exit the Problem

Use this strategy when other efforts have been exhausted, or the vision and strategy of your coach team reveals this client's actions are not a fit.

- Possible ways to exit the problem: the employee-coach is absolutely part of the exit discussion.
- Resist the temptation for the coach manager or sponsor to do it without the coach. The strategic, resilient coach team needs experiences like this to strengthen them with the broad organization.
- Share observations without making people bad and coaches righteous.
- Encourage for future collaboration.
- Always connect the work to the change or transformation vision strategy.

Step 4: Own the Action and Take it, with Support

Employee-coaches should always be involved with the solution to the problem. The employee-coach deserves the opportunity to own the relationship. A manager or sponsor may also want to be involved, and that is great. The coach and leader(s) collaborate together on the communication of the solution. The goal is for all involved to be unified in intent and ownership. Any discomfort felt during the process is short term; the message this unified team activity shows is lasting.

There may be a very rare case where employee safety or some other serious impediment is involved. Those situations are best left to the professionals inside your employee relations team.

Step 5: Reflect for Learning and Close as a Team

When the first four steps are completed, it's time to reflect as a team on the problem, with a lens of learning. A resilient team learns and grows together. Reflection is not a chance to admonish the client or the coach but to learn and perhaps adapt something new to help the team. These questions may help keep your reflection focused and strategic:

- What else can be learned from this problem?
- What else does the team need to move past this problem?
- What else do our management need to move past this problem?
- What can we appreciate about discovering and working through this problem?

Learn from the Loss as a Team

If you're in the change business, you already know many people fear change, and would rather be trained than actually learn something and be transformed. For most, it's the transition *through the change* that is the problem, not the change itself. For example, we want the weight loss or the newly painted room, but we don't want to do the work to get there. We've tried it before and it was hard, and then it didn't work. It's important enough to repeat: most people don't *want* to change.

Relentless messages about vision, and actions of empathy and encouragement help most people move past fear, and more to possibility. It takes time and repetition to get to this point. When people know the shared intent, it's easier to find a way to move forward. Still, organizations and individuals don't know if they are going to see the value of having a coach *of any kind* to help them through the transition to the change.

Often, the coach is blamed for failed relationships and failed transformation. After all, they are the front-line person, and they are the coach with unusual expectations. It doesn't matter how strategic the transformation is presented, or how great of a favorable introduction to the team there was. It doesn't matter if the coach bent over backwards to make connections and build trust. It also doesn't matter if the coach was highly successful in the past. The coach is bad and should be fired or sent to a remote department where they won't be a danger. It's a common event for employee-coach teams, and it's unfortunate.

Management and stakeholders of the coach team need their foundation to lead through these situations: the team's vision, strategy, definition of an employee-coach, and the overarching change or transformation vision. While management usually focuses on the noisy and painful part of helping exonerate the coach, they need to think big. This is hard to do, because the manager is usually getting a lot of pressure to *do something* about that coach, and also to find a replacement coach for the client. Thinking big, back to the reason for the change, and the reason for the employee-coach team offers clarity and usually a next, right move. Keeping the big picture in mind also helps preserve the viability of the employee-coach team, and the big picture of the change or transformation. It's also a great stress test for the team's vision and strategy, and for management.

Leading change is never an easy or clear path. The pressure to conform to existing culture and ways of working can be overwhelming to the small group of those representing change. A strategic and resilient team has the ability to resist the temptation to adjust at every turn, at every squeaky-wheel, emotional manager or team who makes a negative ruckus. Yet, when enough of these difficult events happen, the employee-coach team and management may find themselves at a crossroads. Do we need to pivot? How many situations or events need to occur to consider changing is up to the team and management to decide. When that moment arrives, it's not a time for blame or panic, but one of strategy and retrospect. Make time to evaluate and reflect on what was learned. These questions can help guide that reflection:

Employee-Coach Team Reflection Questions

- What have we learned?
- How are we handling the pressure of being change leaders?
- Does our learning show that we need to make an adjustment?
- Do we have success defined well?
- How might we adjust ourselves to try again?
- How will we inspect these adjustments?

Clean Break, Retrospect Again

There may be enough learning to see that it's time for the employee-coach team to drop a client. The team, and possibly the client, learned *now is not the time* for coaching. This discovery is not failure, and the client isn't bad, although some can be really challenging! The coach might feel failure more than anyone else. Stopping work with a client can be an abrupt event, or one that takes a few days. It should not take longer than three business days to end the relationship. Any additional time spent reveals that the relationship was more than coaching; something for the team to learn and avoid in the future. If it's been decided to make the break, honor the coach(es) involved, the coach team, and the transformation by making an efficient break.

Then, plan to have a coach team and manager retrospective within days of closing the work. This event is aimed at learning for entire team and management. No need to revisit everything that was in the last retrospective about this client; this time is for learning and closure.

Promote Value

Strategic teams promote their value to their clients and potential clients. Promoting the value of coaching helps gain buy-in for more clients, and repeat-clients. Moment by moment, successful coaching creates momentum for the change. When large groups of people begin to understand the value of having an employee-coach, the probability of the team's success increases exponentially. This momentum builds, hopefully engaging hearts and minds in a way that can be sustained.

There is another reason to promote the value of the employee-coach team: survival. When there is a chopping block for budgets and investments, along with shredded cheese on the cafeteria salad bar, some of the first things to get chopped are the investments in training and coaching. The perception of these important activities is typically seen as *nice to have.*

Use a steady drumbeat to promote the value of the team's work. Everyone on the team is responsible to help with this work. *Responsible* is a big word, and it absolutely fits for this work. Neither the employee-coach nor management can keep their head buried in their work, hoping someone else will take care of promotion. The team and management own their success, and so they also own telling their story of success. It is not enough to have enthusiastic clients willing to share the value of their experiences. The team and its stakeholders must continually work to innovate ways to share the value of what they do to the organization.

If you have the luxury of access to an organization change advisor to support your team's successes and events, fantastic. If the change or transformation team has an OCM (organization change management) advisor, consider sharing this person's time with the other team. If these options are not available, ask for small chunks of time, such as two to four weeks, spread out a few times a year. This pulse-like commitment might be more appealing to your investor(s), and a lot can be accomplished in a few weeks of OCM work.

Experiment with energetic, outside help. Seek out an intern from a local university or junior college, maybe even a marketing or branding freelancer looking to get experience. The greatest barrier to these creative ideas is usually the organization's hiring and security protocol.

When you do have support, give them the best chance to learn how to share stories of success and change. Involve them with some of your team's work, so that they can experience the outcomes and successes firsthand. Offer them access to the team and management and get them connected to the change team. Don't make them guess.

If your team doesn't have the luxury of help from an OCM advisor, and none of your change partners have the ability to help you, the team can still promote itself. Remember, everyone is an owner of the team's success story. Lean on the cross-functional teamwork you did during the team's formation: who has talent or energy for this work?

This list of questions can inspire your team to develop a strategy and help inform the next, right step.

Marketing and Promotion Guiding Questions

- What do we want to learn about promoting our work?
- What works now?
- What authentic messages help connect coaching to the change or transformation work?
- What is the intent of sharing our success stories?

- What is the connection to value? As always, connect the value of the coach team's work to the change or transformation vision and strategy, and the overarching vision of the organization.
- Who is our target audience?
- Which stakeholders can help us get the message out?
- How will we reach them?
- Why do people care about the value we deliver?
- How will we measure our promotion efforts?
- Who can commit to helping design this work?
- Who can commit to delivering this work?
- Who can commit to measuring and reporting this work?

Success Stories: Make them Simple and Obvious

Humans who don't want to make the effort to change are not going to work too hard to understand your messages or success stories. Transformation and employee-coach teams often produce success stories that talk over their target audience and end up missing the mark. The team is very close to the work, which can make it hard to share the big picture with compelling messages. It's not impossible to get the message right; keep the mindset of a change leader and let that guide you.

> Humans who don't want to make the effort to change are not going to work too hard to understand your messages or success stories.

These statements from your target audience can be symptoms of missing the mark on your messages:

- *We don't know what this is.*
- *What does this have to do with our organization's strategy and vision?*
- *Why would they do it that way?*
- *Why did they need a coach to do that?*

The problems and opportunities that you coach are probably complex and technical. That doesn't mean your success story has to be. Sharing success, no matter how great or small, from these complex situations needs to be made simple, so that it is obvious. When a success story is obvious, it can be understood by anyone in your organization. Making the complex simple and obvious is elegant.

Balance is needed when considering the length of your message. Think about word choice, and more importantly, the *amount of words used* to make your point. When sharing your messages with those interested in the value delivered by the coach team, more words are not always better. When you find your message is too long, consider if there is too much explanation and not enough leading change. Some explanation is usually needed; don't make assumptions about the client's work or the change happening in the organization. Your audience shouldn't have to work hard to understand any messages. Make value simple and obvious to grasp.

Stress Test Questions for Simple and Obvious

- What is the problem or opportunity, and is it clearly defined in two or less sentences?
- Who is the client and what do they do?
- What is the intent of the coaching engagement/assignment?
- Why is this work important?
- What was the client value from coaching?
- How does this success contribute to the change happening in the organization?
- How does this success contribute to the overall vision and strategy of the organization?

Having a simple and obvious message makes it very difficult to object to it. Value is clearly shown. Some people may not agree with the change, but they will be able to clearly see the what, why, and how the coach team added value. There is no backtracking required for the client or sponsors.

Success stories come in many forms. Try to break out of the usual channels and assumed places with your message so that you are noticed, and, noticed *differently*. Your organization may have a standard way of sharing success; use that method as the foundation for all your messages. Then, push yourselves and management to innovate on other ways to show the value of the coach team's work.

> Success stories come in many forms. Try to break out of the usual channels and assumed places with your message so that you are noticed, and, noticed *differently*.

What you do is important; *how* you do it is even more important. Whatever is created, design it to be a steady drumbeat over time, rather than a splash only when the team has time. Work to design and develop success stories on a cadence and have them available to release when needed. This work may sound big, time consuming, and even overwhelming. It's not as great of a burden as you imagine; it's more strategic than anything else. Be willing to spend the time upfront asking questions to form your strategy and to inspect and adapt the work, and your efforts will be easier in the long run.

This short list of *internal activities* isn't as innovative as it might be inspirational on how to get messages of success and value out to the organization:

- Webpage with stories, testimonies, supporting data, such as definition of coaching at your organization, and proposed coaching activities.
- Presentation that any team member or manager can deliver to potential teams, management, communities of practice, or other internal groups.
- Blogs
- Podcasts
- Videos
- Photo collages

- Webinars
- Management acts as relentless storytellers at high profile meetings

Look Beyond Your Organization

Yes, there are company regulations, guidelines, and other restrictions to honor and uphold for any community activity. This handbook assumes all organizational guidance is accepted and followed (or formally challenged!) with the right stakeholders in the room.

Honor the restrictions but don't let them hold your team back from innovation. Strategic, resilient coach teams look beyond their own organization to gain perspective and to learn. Your coach team is unique, yet there are likely others in the industry who are similar to you. Even better, meet with those very different from your coaching team, and be open to unexpected learning.

> Strategic, resilient coach teams look beyond their own organization to gain perspective and to learn.

Rely on your team's vision, and marketing strategy to shape your outside activities. Then, craft your intent. Keep it simple: refine if your intent is to give value to the community, get value from the community, or both. This simple strategy provides great focus and avoids wasted time, energy, and money. This list can potentially fuel your team's imagination to develop a great list of activities:

Ideas for External Community Connection and Learning

- Participate in local meetups or groups from your industry, or adjacent to yours
- Visit other coach teams, virtually or in person
- Write and publish a white paper

- Publish an outside-facing blog
- Submit to speak or give a tutorial at a conference
- Be a guest on an industry podcast
- Host a new or existing meetup or group meeting
- Create a new conference or *unconference*
- Find an unlikely organization and partner with them on an event
- Partner with a charity that connects to your coach team's work

Hire a Coach of Coaches

The best must get better, no matter where they are on the journey. Strategic, resilient employee-coach teams use an outside coach to support their deliberately developmental strategy. These teams see the value in continually learning and improving themselves for the client, the transformation, and for personal growth. The team's stakeholders work to ensure the team gets the coaching they need to be successful. In past team strategy sessions, when the questions, *"How will we learn?"* and, *"How will we be successful?"* were discussed, having a coach of coaches was hopefully one of the ideas. The needs of the coach team drives what kind of coaching is needed, for how long, and how often.

For example, an established employee-coach team of six people had two coach of coaches when they first formed the team. For eight months, they received coaching and support on basic coaching concepts, team building, and then created a plan for continued growth and development after the contract ended. It's been almost two years since the employee-coach team was established, and they are looking to get to a new level on their growth plan. There is a lot of corporate noise and many distractions, making it difficult for the coaches to keep their focus on the key parts of getting themselves to the next level of their development. They know that to properly support the needs of the ever-changing transformation, they must grow faster than they can do on their own.

Next level may be a new term, or perhaps an over-used term in your world. For the guidance offered here, next level is a real concept. The list below provides examples of what this book intends by using the concept:

Examples of a Team's Next Level

- The transformation has pivoted, and the coach team needs enhanced skills to support it.
- Need for the entire team to gain new knowledge, and then how to coach to it.
- Desire for the team to be more coach and consultant-like, less like order takers.
- Many of the original team members have left, including some or all management.
- Need for more resilience; less drama within the team and with clients, more bulletproof to distractions.

Many other next level situations exist; the bottom line is that what got you here may not get you there. Having a coach of coaches creates problems and opportunities very similar to those experienced as an employee-coach, such as building trust, making time for coaching, and being open to change. The best way to ensure your team gets the value they deserve from a coach of coaches is preparation…for the team, management, and coach. This section highlights some of the key points worth repeating.

How to get the value your team deserves from a coach of coaches:

Establish Clear Outcomes

Make the desired outcomes of the coach of coaches clear to all involved. Revisit them often, to make sure the engagement is on track. For example, if the team has a retrospective every two weeks, review the coach of coaches outcomes once a month. Humans are dynamic and changing, so expect the outcomes to slightly shift and adapt over time.

Be Prepared to be Coached

The employee-coach team needs to be ready for whatever is included in having a coach of coaches. *Ready* includes a willing mindset to continuously learn, and the time to participate in most or all coaching activities. Examples of *ready* include making the time needed for coaching, collaboration, shadowing, pairing, and homework assignments. Homework could include book clubs, research, writing or speaking assignments.

Be Honest about Growth and Development

Don't fake learning. A deliberately developmental team is open and honest about what they *want* to learn and what they *need* to learn. Having a coach of coaches is not a sign of weakness but one of strength and commitment to continuous improvement. The more clarity the team and coach of coaches has about what they need, the more strategic the relationship and outcomes.

Own it!

When the team owns the coach of coaches relationship, they participate in what is asked of them, so that they can get the most benefit. They ask questions, are curious, and challenge the coach of coaches in order to deeply learn. When things don't make sense or are not going well, the team speaks up and has an honest discussion about what is happening.

Experiment

The employee-coach team understands the value of small-scale experiments. The team has a mindset of experimentation to learn, rather than all-or-nothing-thinking in dramatic form. Even if the team has a best-known way or method, they keep an open mind to trying new approaches offered by the coach of coaches.

Have some Perspective

Having a curious mindset willing to learn and experiment also means an openness to other perspectives. If the employee-coach team is lucky to have more than one coach of coaches, they should actively seek out perspectives from all of them, to gain the most insight and experience. This situation is usually a short time in your coaching career; make the most of it in all ways possible. Negotiate with management, if you have to, so that you get every drop of value you can.

Curiously embrace diversity of thought and coaching practices. Most coaches will not run you off the road; their advice and methods are usually in a wide spectrum of *right* that is open to interpretation and perspective. Disliking someone's advice does not make the advice (or the person who gave it) terrible.

When these coach of coaches offer differing thoughts, they are not considered right or wrong, but different. When the team picks one approach or method over another, the other methods, or the people who created them, are not bad. Only in extremely rare instances does a coach of coaches offer technically terrible advice to a team.

Stay Connected!

The employee-coach team and management know they have a coach of coaches for a finite amount of time, so they spend as much time as they can with them. The team brings the coach of coaches close to their work, knowing that the more the coach can see into their work the better the outcomes. The team knows the coach is not a team *member* but is embedded in the team.

Have Fun

Strategic and resilient employee-coach teams know how to have fun, and they can do it with their coach of coaches, too. Learning and getting constant feedback is hard work and deserves some levity. Having fun also helps build trust. A coach of coaches knows how to time their professional humor, so find a way to make learning enjoyable.

Learn as Much as You Can

The curiosity mindset helps the employee-coach team be relentlessly vulnerable in asking questions. They know the coach of coaches is dedicated to their continuous improvement, so they spend time in group and individual interactions to learn as much as they can from the coach.

Know When to say When, Again

The employee-coach team knows they need help, and when they must try it on their own. From the first day they work with and learn from the coach of coaches, the team is aiming to work independently of the coach. They also know they may need a coach of coaches again someday, when they need to achieve another level of knowledge or expertise.

The other side of the coach team knowing when to say when is about missing outcomes. In this case, the team and management have done enough retrospective to discern between disagreements in perspective and a failure to make progress toward outcomes. It's time to try a new coach of coaches.

Selecting a Coach of Coaches

All's well that begins well. Normally, employee-coach teams are very different than every other part of the organization they serve. The employee-coach role is intentionally different if it is to be successful. And, the coach(es) who develop and grow the employee-coach team can't be successful if they are selected the same way as everyone else. The phrase, *"We're different"* is usually used as an excuse for avoiding change; in the case of a coach of coaches, the difference is legitimate and healthy.

Because the work is different, involve your *partners in change,* usually the change and transformation team. Then, use these ideas to inspire your search for your team's coach of coaches:

Take a close look at who is interviewing the coach of coaches, who needs to be a seasoned coach *and* consultant. Don't leave the hiring to someone who simply knows the process of hiring people in your organization. The hiring person absolutely needs to be dialed in to the employee-coach team and your partners in change. Employee-coach team members can be involved in the interview process; two other neutral people should also be involved.

Stay grounded in the vision and strategy of the employee-coach team and the transformation. When personalities and skills are even, how a candidate holds up to the vision and strategy can guide difficult decisions.

Get a sense for the candidate's coaching philosophy, not just knowledge and tool use. Beyond experience coaching other coaches and referrals, search for the candidate's published work, conference talks, videos or other media. Think outside the box. What does this statement mean for your team regarding hiring a coach of coaches? Working with one?

Aim for diversity of thought. Don't select people like you, in hopes of being better understood and easy to work with. Diverse thought gives your team the best chance to broaden their perspective and gain experience they wouldn't get any other way.

Practice Resilience

A resilient team knows how to bounce and learn without falling apart. In a rare moment where a resilient team might fall apart, they regroup quickly, and learn from the experience. Resilient teams have a mindset that is hungry to learn how to improve themselves and what they deliver.

This handbook provides the framework and practices to begin growing a resilient team, no matter how long they have been together. Writing and talking about resilience is easy to do; actually building it in a team is hard. It takes an investment of time at the beginning to strategically establish the team, and then relentless follow through.

With a strong foundation of strategy and vision, a deliberately developmental environment for the team and management, resilience is possible. One moment at a time, a team and its management become stronger and less likely to break when things get difficult. This section offers guidance for some of the most common growth moments that your team will likely encounter.

Avoid All-or-Nothing Thinking

Resilient teams know there is a future beyond any problem, even when it feels like there is none. Present moment problems are solved with the future (strategy and vision) as a guide. Also, a team with a continuous improvement and developmental mindset can make their work...smaller. Coaching is important and impactful to the organization, but it is not a stagnant, one-way highway. The resilient coach looks for multiple options, even when it feels like there is only one option. Remember the *"I had no choice"* statement in the Victim - Coach Failure Patterns.

Feeling Left Out

Change and transformation work is not as strategic as it could be. People feel many things, especially excluded. *"Why weren't we told? We always hear about it from our clients first!"* There will be trouble with inconsistent change leader and stakeholder communication and activities. A resilient team will always work to expect the best of their change partners and clients and knows there will be many imperfections. It will feel intentional, but these inconsistencies and undulations are not usually meant to attack or ignore the employee-coach team. The resilient employee-coach team looks at transformation and client missteps with curiosity rather than indignation.

What to do When You are Called on the Carpet

Along with the roller coaster experience of coaching a transformation, the employee-coach team will also find themselves being called out for some kind of objectionable behavior. The coach role and the transformation are often misunderstood, which leads to labeling those involved as bad, ineffective, even insubordinate. The strategic and resilient employee-coach team has a lightweight, visible process to investigate the problem while supporting their coach(es) and the transformation effort. The process does not include assumptions about anything, secrets, or rumor spreads. With a learning mindset, the manager and the coach involved resist cultural and peer pressures to remain focused on openly getting to the root cause of the problem. The resilient employee-coach team sees past the discomfort of learning from a client experience to apply the learning to the future.

Attachment

Resilient coach teams are always supporting the change or transformation using curiosity and a learning mindset so that they do not get attached to a single outcome. The undulating effects of their own coaching work and that of their change partners is far less disruptive with an expectation that change is constant. The finesse of not being attached is in having a learning mindset.

Be Generous

Resilient employee-coach teams are generous with what they learn with each other, the organization, and possibly the global community of their coaching specialty. The team and management know that they are not losing anything by being generous with their time and their learning. There is only upside to a team becoming a generous, thought-leading entity.

Hunger

Resilient teams know the value of stopping, even just for a few moments, to reflect and appreciate what they have learned. They celebrate their wins, their learning, and the even uncertainty of what's next. They talk about change, and then actually follow through. Their managers encourage them to be hungry and reach for more, as individuals and as a team.

Part II
The Individual Employee-Coach

Part II provides the recommended strategy and approach for personally growing into a strategic and resilient employee-coach.

The Employee-Coach

Using the guidance in this book, your team has the basics to establish or re-establish itself as a strategic team that can grow its resilience over time. There is one part missing from this guidance: you. The whole you, not just the knowledge and experience parts of you.

Similar to Part I, we will use the constructs of operating with strategy, this time focusing on the individual coach. When growth and success are the goal, the employee-coach has a lot of individual work to do. By the way, if you haven't done much work in this area lately, now is a really good time to get back into the habit. There are so many ways to grow and sharpen your knowledge that most of your competition is doing it in some form. Why not join them? Perhaps even lead your own growth tribe.

Individual work begins with operating your coaching practice with strategy, reflection, and a growth mindset. Being part of a team helps shape the kind of coach you are, but it is not the only way to grow your practice. And, it is *practice.* No coach ever has all the answers, and rarely does the same problem or opportunity happen exactly as the last time.

Individual growth of your coaching practice will ultimately help you be successful and far more resilient than you ever imagined. Being an employee-coach may sometimes feel limited, even stagnant. When you operate your own coaching practice with strategy, you will feel less trapped and more open to discovery and growth in the face of restrictions. When you *feel* less trapped, you *become* less trapped.

Chapter 4
The Business of You: Foundation

The business of you **includes three important topics:**

- Foundation
- Mindset
- Follow through

If you are an employee-coach or about to become one, establishing the *business of you* can no longer wait. Now is the time. In fact, now is the perfect time to commit to your personal development. That is the *what*. *How* you live out your personal development is wide open to interpretation and possibility. There are so many ways to grow and explore your potential. What is offered here is not the only way to do it. The intent of Part II is to give you ideas and inspiration for creating your career development (notice I left out the word *plan*) and a little structure for the *how*. You will see this work is similar to that in the team-focused Part I, Chapter 3: Operate with Strategy. For this section, keep your mindset on *your own* growth, and enjoy this generously narrow focus, just for you.

Forget the career development plans, 360 feedback, that list conferences you plan to attend, or what certification renewal is next for you. Also forget about your performance achievement plan. Some of us would like to completely throw out performance achievement, but let's leave that for another discussion. Those plans and action plans can be good tactical things to develop and aim for in the near and short term. There is more long-term, introspective work to do first. You've probably heard it before, maybe even did some of this type of work in the past. But then something more distracting came along, and this work was left for later. Much later.

True career development includes a personal vision, a supporting strategy, tactics, and a whole lot of personal reflection, and sharing your journey with someone else. The mindset for career development includes curiosity, continuous learning and improvement, and growth. Sounds a lot like the mindset of a strategic and resilient employee-coach team, doesn't it? Career development is a process that unfolds over time but does not require as much time as many imagine.

If you have practiced true career development in the past, but it's been a while, try to see your work with fresh eyes. Resist the temptation to simply assume you can pick up where you left off. Read this section, do some personal reflection, and perhaps discuss with your coach of coaches.

Personal Vision, Strategy & Vision Board

Personal vision is often misunderstood, which leads to misguidance and missed growth. Your personal vision is a statement of who you are *becoming*. It is a short, future-focused, well-defined statement about who you want to become. What is trying to unfold and happen inside you? It's not a feel-good statement about who you are, your personal values, or what you have done so far; it's future focused. Think really big, here, beyond certifications, achievements, and education.

Benefits

Why have a personal vision now, when you've made your way in life seemingly very well without one? *The business of you* needs grounding and purpose. When things get difficult, and you know they will, you will have a foundation to base your actions. A personal vision serves the good times well, too. You may have a fantastic opportunity that seems almost too good to be true. A personal vision and the supporting practice of reflection will help provide the clarity needed to act. Or not.

If you've tried a personal vision in the past, and it's been a while, consider starting again, using the guidance here. Don't pick up the historical work;

look forward and begin fresh. Whatever worked or didn't work in the past is learning; your personal growth is worth another iteration.

Depending on the type of coach you are, having a personal vision may be part of modeling what you are asking of others. Your own experience of creating your personal vision and a supporting reflection practice will enhance your coaching abilities and credibility. You don't have to share your vision with your clients to share the value of your experience. Yet, if you are comfortable and the time is right, lead by example and share it.

How to Grow: Guiding Questions

Use reflection and these questions to help shape your personal vision. Potential answers are in italics to fuel your ideas.

- Who am I?
  I am a daughter, a Mother, a sister.
- Who do I want to become?
  I want to become a servant leader in my family and work, and be financially secure.
- Am I ready to change? Maybe some things have to happen first.
  I am ready to change because I am so tired of where I am now.
- What might I do to be ready?
  Make time to regularly reflect. Ask my coach for help.
- What must stop for me to become this?
  I must stop thinking I don't have time for this work!
- What must begin for me to become this?
  I must book time in my calendar, so I never forget about it.
- How can I express or show who I am becoming?
  Make a vision board. Share it with my coach.

Examples of a Personal Vision Statement

- My heart grounds me in my personal growth, and I am close to people who believe I matter.
- I am free of debt and unconscious decisions.
- I am a servant leader who mindfully grows and makes a positive impact on my family and community.
- I am a leader who is curious and generous in my thinking and actions.
- I am healed and unattached to the past.
- I am mindful and intentional with my heart and mind.
- I love myself and I love my life.

It Will Feel Different

Don't worry if you haven't done this work before; now is the perfect time to create and use a personal vision. And, it's really important to understand these statements aren't created in one reflection session where everything magically falls into place. They are created iteratively, over time. Expect to feel bad before feeling better. You may have to go to places you have avoided or dislike, which is why many people never finish their personal vision work. Why drag yourself through your own mud and dirt (again) if you don't have to? Because personal growth demands it. This part can get really deep, requiring the help of a coach or even a therapist. Getting help to understand what is needed to grow or get unstuck from a persistent muddy spot in your life is not failure. The act alone of seeking the conversation with someone else is growth. A Jedi builds their own lightsaber, to learn, understand, and use.

> Expect to feel bad before feeling better. You may have to go to places you have avoided or dislike, which is why many people never finish their personal vision work.

So, if you are willing to do the introspective work, you now have some writing work to do. Succinctly writing who you are *becoming* is hard work. Yet, it's your journey of growth, so the only thing that can be wrong about it is if you don't do it. Experiment; find what works for you, and don't be attached to a single approach. You are a dynamic human, with different moods and different attitudes. For example, some days of the week and times of day may work better than others. If you normally work on your vision on Wednesday evenings, but a day packed full of online meetings socked your energy, don't try to grind through it. *Becoming* is different, remember?

Your vision statement may serve you for a short time, or for the rest of your life. Through the practice of personal reflection over time, you will notice when it's time for an adjustment to your vision. Working with a coach or mentor can also help you see the big picture and the nuance of gentle nudges for change, or to stay the course.

How to Express Your Vision

Writing your personal vision is a perfect way to express it. So why are we still talking about this? Because in addition to physically writing your personal vision, and practicing the reflection to support it, expressing your vision another way helps deepen and sustain who you are *becoming*.

Reflection can also be enhanced with an expression of your vision that resonates with you. Making your vision visual and relevant to you is key. Creating a vision board is any easy way to express your vision with images. Before you roll your eyes thinking of social media feel-good boards or bucket lists, pause. Remember this work is about your personal vision, about who you are *becoming*. A vision board is absolutely different than most boards you will find in media. A vision board is a tool, and it is a practice. The tool part is in making the physical board, and the practice is in reflection and iterating on your personal vision.

Your visual representation of your personal vision could be in whatever form helps you keep *who you are becoming* regularly seen by you to uphold its importance in your life. The words in your vision are represented with an image or drawing, maybe even a sculpture of building blocks or some other medium you enjoy.

Types of Vision Boards

- Small board, using an letter sized sheet of paper
- Large board, using a giant sticky note or poster board
- Lego or clay sculpture
- Digital version is ok, provided you make a physical version of it and make it visible

Guidance for Making a Vision Board

I use the words vision board in this section to cover all of the expressions of a personal vision.

- Your vision board is a strategic tool, used to supplement your practice of reflection, and the nurturing of your personal vision. The intent of a vision board is not to simply feel good or create something to invite magical interaction from the universe.
- Personal vision work is not a one-and-done event, rather an ongoing experience that you nurture over time.
- Just like starting a fitness program, or even a new job, it takes time and experience to know if you're on track. Be assured that anything worth doing is worth doing well, so give your personal vision practice an honest effort.
- You don't need to be an artist to make a meaningful vision board. This is a personal tool, not a work of art in the formal sense. Try creating your own, even if you don't like how it looks initially. If that doesn't work, find what works for you. If it's dried pasta and string, so be it. As long as you feel its relevance to your vision.

- You don't need an image for every word in your vision. Your board should visu-ally remind you and inspire you of your personal vision. When you look at it, you have a feeling of connection to who you are becoming.
- It's acceptable to have a few words written on your board. The goal is to use images whenever possible, but sometimes one word can say it best, such as *heal*, or *breathe*. Those were actual words on my vision board, when I didn't feel a drawing or image could properly represent it.
- Find a visible home for your board where you live or work. Or both. There are usually plenty of options in your living space that don't involve hijacking the kitchen walls or the wall with the family photo gallery. Having a board at work may be the same as where you live, but if you work in a different location than your home, you could post your board or a mini version of it at work. Don't feel obligated to do it, as your personal vision is truly personal.
- If a vision sculpture, or some other physical expression of your vision works better for you. All of the same guidance here applies.
- Careful with self-talk about your board; own it as the best representation of who you are becoming. Avoid criticizing what you created. It's a tool, not an evaluation. Keep your words and thoughts about your vision board positive and energetic.
- A personal vision board is a strategic tool that supports you best when you nurture it over time. Who you are becoming won't happen overnight, and will continually change with your growth.
- There will never be a perfect time to begin making your personal vision, or the board to express it. *Now* is perfect. And, now is also a perfect time to start working on your board.

Your Personal Strategy & Tactics

You've got your personal vision, who you are *becoming*, but how will you even begin to get there? Start by making a personal strategy, which is a statement for *how* you will work to reach your goals. This statement is not meant to capture everything you will do, but the essence of your *how*. From here, you can develop the actions, known as tactics. There can be a lot of tactics to support a strategy; try not to over-engineer the list. The samples below are lightweight and incomplete, but long enough to demonstrate how to make strategy and tactics that support a vision.

Sample Personal Strategy Statements and Tactics

Strategy: I learn each day with focus and empathy.
Tactics:

- Read at least one book a month.
- Make time to ask others for their perspectives.
- Learn at least one thing each day. Journal it.

Strategy: Grow my Coaching Practice through Relentless Learning and Experimentation
Tactics:

- Approach situations with the mindset of curiosity.
- Attend methodology conference.
- Share and get feedback on my experiments.

Strategy: My growth is based on curiosity, practice, and courage.
Tactics:

- Define the learning experiences I want to have.
- Work with my coach of coaches for two hours each week on my desired learning experiences.
- At least once a week, share my thoughts at a time when I normally would not.

Strategy and tactics are smaller and more defined, so that they can adjust and adapt as you live out who you are becoming. Personal reflection is the best tool to refine your vision and to inform any adjustments to your personal strategy or tactics.

Personal Reflection

It's life-changing work, if you are willing. Reflection and search for who you are *becoming* is a powerful tool for personal growth, and it's absolutely free. It's connected to, but different from, meditation, which is another rich topic worth exploring after this book. Using reflection, and the focus you gain from it, will help you improve your view of the big picture. Your personal vision needs continual reflection to refine it and to live it out. The desire for who you are *becoming* must be strong enough to withstand the pressures of short-term distractions.

Most of us are constantly pulled to the present, to the near-term, noisy needs of right now. These things require your attention; but if you are not able to step away from them, you could wind up a reactionary slave to your surroundings. There's more to you than what is happening in the short term. Reflection helps keep your vision at the center of all you do.

The stacks and stacks of decisions you make all day long, every day, are micro decisions. These daily, often unnoticed decisions add up over time. With reflection and focus, they will add up in a positive and creative way. We're human and imperfect; sometimes we'll make the wrong choice or make choices that are only mediocre for our growth. The regular practice of personal reflection helps you notice patterns and trends in yourself; nurturing those you want to continue to grow and stopping those that hold you back.

Reflection needs space. Mental space that can't be accessed when there are distractions. Packed days with no space to breathe creates a stagnant pattern in your mind and body. Some days and weeks will definitely be challenging; it's when those few days begin to be a pattern of living and working, devouring any other activity that stands in the way. If this is you, there is hope. Like any change, it takes commitment and practice to regain balance.

If you've practiced personal reflection in the past but something disrupted your practice, now is the time to reconnect to it. Keep an open mind about it, even though you've practiced in the past. Maybe the same practice you used then will work for you now; maybe it's time to try a new approach.

How do you start giving yourself space? Take really small steps. Begin by giving yourself screen-free space for just five minutes today. Try it for five straight days, if you can. See how it felt to break away. Assuming you could break away for five minutes, how did you feel during that five minutes? How did you feel after?

Does this sound like a big ask? Like an enormous shift in your 24-hour day? If yes, then you have too much going on (WIP) to improve yourself. That is the bottom line. No judgment, here; only an observation. Just like a team needs to make time to regularly reflect to grow and reach its potential, you need to as well. Without regular, personal reflection, don't expect to improve or grow to your full potential. There is hope to change this mindset, and since you are a coach of some kind, there are people around you who can help.

Ideally, your experiment to reflect for five minutes for five days was valuable to you. And, that first five minutes without any screens might have felt like forever. Similar to going to the gym or starting any new habit, immediate results rarely happen. Don't expect a giant event where the sky opens up and you suddenly have all the answers. Change will happen in small increments, over time. Expect a time-released opening of your heart and mind, with space to explore thoughts and emotions. It's about nuance. Most of the obvious problems have been solved, remember? Your contributions to noticing nuance, to the details that matter to your growth and life purpose, are far more important than checking your news feed or email.

Examples of Personal Reflection Approaches

Meditation

Meditation makes the space for personal reflection. Be curious about using it. Meditation can help clear your mind of the past and future, to open your mind and heart to what is happening in the present. With an open mind and heart, it's much easier to do personal reflection. Meditation is a big topic, with many options to explore and get started, or re-start an abandoned practice. To keep it simple, there are meditations done in stillness, and those done in movement. For example, a moving meditation could be a silent walk, run, or any movement outside, without a device of any kind. There are all kinds of meditation apps and music. I recommend trying those only after you've tried meditation without any devices for a few months. Stillness meditation is also device-free, and silent, preferably without other people to distract you. Make yourself comfortable, even if you're only planning to meditate for a few minutes.

Reflection Journal

A more formal way to reflect, a reflection journal offers a way to actively explore your thoughts and emotions. Reflection journals have also been known to enhance critical thinking skills, knowledge, and improve writing skills. All upside for you, with little investment. You don't need a mermaid-sequined journal and fountain pen, although if those items help sustain your reflection practice, please enjoy them. How often you journal is up to you; I recommend once a week to get started. As your coaching career undulates and varies in intensity, it might be valuable to write a reflection each day.

Write your vision and strategy on the outside or inside of the journal. If you have a vision board, consider pasting a miniature version of it on your journal as well. You will see this visual connection each time you reach for your journal, which will reinforce the big picture, and the *business of you.*

When you review and reflect on what you wrote in the past, do so with curiosity and kindness toward yourself. It's so easy to read a past reflection and say, "*Wow, I really had no clue what I was doing!*" The point is not to judge, but to curiously reflect. That's when growth has a chance.

Share Your Reflections

Another effective approach is to share your reflections with another person, on a regular frequency. The person you share with is ideally your own coach, or perhaps mentor. The right person for this job should be able to listen deeply, be objective, and be willing to curiously explore your reflections. It can be difficult to find a person with these characteristics, who is also someone you can trust with your personal growth. Seek them anyway.

> It can be difficult to find a person with these characteristics, who is also someone you can trust with your personal growth. Seek them anyway.

Simple Structure for Shared Reflection:

1. This week, I observed...
2. This week, I learned...
3. This week, I am reading...
4. My current experiment is...

The beauty of these questions is that they are strategic and yet open ended. The first time I used them was actually in a monthly meeting with my manager, called a Monthly 1x1. The kind of meeting that was incessantly moved around the calendar, and chronically changed from 60 minutes to 30, sometimes even 15 minutes. These meetings were so awkward because the well-intended manager spent much of the time talking and I sort of sat there, thinking about lunch. I had long given up on trying to work in a word, a thought, or an interruption. Each time I tried; the meeting always reverted to the manager's control. This meeting was just a Monthly Manager Monologue. I couldn't avoid the meeting, so I decided to restructure it, without announcing it to the manager. I simply used the highlights of what came out of my personal reflection to share in the meeting. I started our meeting with the statement, "I have four reflections to share." This simple statement, including a number in it, permanently changed the bias in the meeting to strategic and reflective. One time I forgot to mention it at the beginning of the Monthly 1x1, and the manager asked me, "Don't you have your reflections this time?" It was a great experience for me to pull the answers to these questions from my personal reflection time, and to prepare for that meeting, rather than dread it. I don't recommend you must share your reflection with management; this example was only to show you the strategy and power of personal reflection.

Chapter 5
The Business of You: Mindset

The *business of you* needs a strong foundation and a mindset to support it. Successful coaches build a business owner mindset over time through practice, determination, and experiments. Even though they are an employee of a company, and must work in many systems, they can see themselves as an entrepreneur. Use the following guidance to inspire and grow *the business of you* mindset.

Experiment

Avoid all-or-nothing thinking. As a swim meet announcer for my daughter's swim team, I have witnessed many fantastic moments. In 2019, I announced a championship meet that had 230 female swimmers compete in the 50-meter freestyle event. There were 12 ties (exactly the same finish times) in the event, and in one of the heats, three swimmers had the exact same time. Yet the effort for each of these swims was unique. Different body types, different dives into the water at the start, and nuances of stroke technique. There is more than one way to do it right. Knowing this fact helps us feel more comfortable with experimenting and testing a hypothesis, so you can have a looser grip on your ideas.

A looser grip doesn't mean not caring or not having focus; it means believing your growth consists of *experiments.* This mindset helps you become more resilient over time, which will make you stronger, and enhance personal growth. You're less attached to your activities and their outcomes. When things don't go well or they fail, it's learning. Failure is learning. Knowing what does not work is powerful learning. Failures and misses can still knock you down and hurt, but the looser grip helps you let go faster, and look ahead again.

> Knowing what does not work is powerful learning. Failures and misses can still knock you down and hurt, but the looser grip helps you let go faster, and look ahead again.

Growth

Expect growth to happen in leaps, and also in nuance. This approach doesn't mean you are lowering any expectations of progress. Expecting change through nuance is actually a more advanced and mindful approach to growth. If you've been reading along, you understand the basics of nuance and coaching. Remember micro decisions, the small ones you make all day long? They are nuance. And, they often go unnoticed, which means they are unaccounted for. They pile up. Until one day, they are noticeable. Apply your vision, and practice reflection to give focus to your days, so your micro decisions add up in the most positive and creative way. Sharing your growth with someone else gives you a higher chance of discovering nuance, and enhancing your journey of growth.

Truths about Employee-coach Career Development

Unless they are an experienced, professional coach, management usually does not know how to develop employee-coaches. With much respect for the people in these roles, their organizations usually do not prepare them for guiding personal development. The well-defined path of a company's career development and performance management processes are focused on plans, goals, and tactics rather than helping someone create a personal vision and strategy. Your actual career development is not a platform to improve any process but to focus on your growth. Yet management are usually experts in process and systems that help evaluate you as an employee. Honor them, and the system they represent.

Career Development of the Employee-coach is the First Thing to be Dropped when there are Challenges and Distractions

After all, you were hired to coach, right? Similar to the organization investing in training and coaching, career development activities are often the first to be cut when there are distractions. In the case of career development, it's not always about money, but time and focus. Lost commitment usually starts with a one-hour career development meeting being rescheduled multiple times. Then, the meeting time is in the scraps of the day, and likely, the week. Can you make Friday at 4:00 pm? How about we shorten it to 30 minutes, and schedule it on Tuesday at 7:00 am? You'll show up for the 7:00 am meeting, only to get a cancelation at 7:05 am. Foiled again! The scraps of a day or week are the times on the edges of the organization's typical workday and week. Most people use this time for starting or closing the day, including getting unfinished work to done. So, when a career development meeting is constantly shortened, moved around the calendar several times and/or lands in the scraps of the day or the week, it sends a message that the meeting is not important.

It's up to you to own your career; what can you do about a lack of commitment to your career development? With an open mind, a lot.
Some potential ideas:

- Keep going to the same well, trying to get time with your manager, and hope for the best.
- Negotiate with your manager for more regular time and the full time you need.
- Find a mentor outside your division, but within the organization.
- Find a mentor in the greater community of coaches in which you belong.
- Ask your greater community of coaches for help, open source style.
- Ask your coach of coaches for help.
- Hire a coach.

Finding the Right Person to Help you Grow is Difficult

The right person is not simply a manager or coach willing to offer you true career development; they also need to be a good fit for your personality and style. This person may or may not think like you; but they are someone you can respect and listen to. Lucky you if this person is also your manager. If your manager is not a good fit, you have options listed in the prior section.

True Career Development Might not be Taken Seriously by Others

Not everyone, including your leader, are motivated to grow as much, as deep, or in the way that you do. Not only will you need to persistently advocate for yourself to get career development time, and find the right person, you will need *willingness.* When done right, career development is undulating work that can feel great and uncomfortable, all at once. Success requires your willingness to *go there*. To do the work, the experiments, push through or past the discomfort of unknowns to learn and grow, with the future in mind. Few employee-coaches are willing to *go there*, and even fewer managers or coaches are willing to do the work to help you go there. Outside help is your best option.

> When done right, career development is undulating work that can feel great and uncomfortable, all at once. Success requires your willingness to *go there*.

Work in Small Chunks so you Don't get Overwhelmed

Keep your focus small and be persistent. The most impactful work supporting your vision and journey is small and constant.This is the nuanced work that adds up slowly, over time. The small things are easier to adjust, too.

What Got You Here Won't Get you There

Your coach of coaches in the past may have been fantastic for helping you grow. That fact will always be true; it does not mean that this person is right for your next level of growth. With a curious and experimental mindset, work with your coach of coaches on your next growth target, and then use personal reflection to validate that this person is really helping you.

No One Cares About *the Business of You* More than You

It's your vision, and yours alone. You own who you are becoming. Even the best manager, coach or mentor who helps with your career development doesn't care about it like you do, so don't expect them to. They have ownership of the commitment to help; you have ownership of creating your path, and then living out the work of your development. Notice I didn't say deliver or execute. Personal vision and career development are about *becoming*. Make it something that the rest of your life cannot ignore.

Get Involved in the Community

One of the best free options to grow *the business of you* is to engage the greater community. When you look beyond learning from your organization and your coach of coaches, you have a wide world of options. Your personal vision will help you narrow it down a bit, but don't be surprised if you find an overwhelming amount of learning to do with others. Collaboration opportunities, meetups, conferences, and volunteer work are some of the obvious opportunities. There truly is something for every interest and energy level, if you are willing. The reason there are so many opportunities to learn in the community is because there are many people doing the same thing as you: looking to grow themselves.

It might feel like having a new, very challenging job should be enough to take on. After all, having too much work in process is overwhelming and will slow you down and scatter your vision in the wind. It will always feel like there is not enough time to go beyond the expected and usual work. There will always be something more important than your personal growth and development. That's why you have a personal vision and strategy, to push you past the ordinary and expected. The best coaches of any trade are active in their greater community in several ways.

Like most other successful coach activities, take an inspect and adapt approach to how you engage the greater community. Find what works, given the amount of time and energy you have. Try pairing with a motivated coworker, if you need the push. This coworker doesn't need to be a coach, either. This person doesn't need to know your personal vision to be a great collaboration partner. *Be willing.*

> Like most other successful coach activities, take an inspect and adapt approach to how you engage the greater community.

Chapter 6
Follow Through

You've got your personal vision, maybe a vision board, a strategy with tactics to carry out the vision, and a budding reflection practice. You're on your way. You might ask, *"On my way to what?"* On your way to *becoming* who you want to become. You've created the foundation to grow and become. Now you must establish a *practice of becoming.*

This is the point where many people stop: when it's time to follow through. Depending on the type of coach you are, you may know this human behavior very well. Just because a coach is aware of this human condition doesn't make it any easier to overcome it. People want the change, will set themselves up for it with great intentions, but they can't make the transition, because they don't do the work. So the practice of *becoming* stops. The daily and weekly habits of personal growth are not made the priority that they could be, and the practice is soon forgotten. Sometimes, there is an honest disruption to the work that then makes it challenging to restart the practice of *becoming*. These interruptions are less common than you might think. As coaches, we know that life in general is a series of distractions and challenges that never lets up. Having a personal vision and strategy keeps focus in a world demanding your attention for everything. The vision and strategy you created help to bullet proof the work of who you are *becoming*. What you created is strong enough to withstand most any distraction, and flexible to adjust from what you learn over time. But it is nothing without the practice. Relentless, curious practice that changes over time.

> As coaches, we know that life in general is a series of distractions and challenges that never lets up. Having a personal vision and strategy keeps focus in a world demanding your attention for everything.

Learn, Adjust, and Begin Again

If you fall out of practice, don't be hard on yourself. Learn from it. Losing focus happens to the best of us, even with all our good intentions. The key is to quickly restart again without blame or negative self-talk. Learn, adjust, and begin again. When you practice this, you are becoming more resilient in all parts of your life. Reflect on what happened and learn about what stopped or slowed your practice. Take what you learned, and then begin again with a refreshed approach and new energy. *Be willing.*

> If you fall out of practice, don't be hard on yourself. Learn from it. Losing focus happens to the best of us, even with all our good intentions. The key is to quickly restart again without blame or negative self-talk.

If you're not sure you can do this work on your own or if you have been idle or stuck, consider a personal coach to get started, or to refresh your practice. A personal coach can work with you to establish your foundation and then push you to follow through. Having someone in a structure to hold you accountable to the practice of *becoming* may be uncomfortable, for all kinds of reasons. Are you willing to go there? It's those people who are willing to accept the push in accountability who will find the most success in a coaching relationship. Speaking of discomfort, a coach will teach you to progress forward *especially* when you are uncomfortable. A good coach will teach you to feel comfortable in the discomfort. This is the work of *becoming*. The outcome of that work is that you feel right at home with your vision and ambitions, when many in the world would rather enjoy the comfort of complacency.

> A good coach will teach you to feel comfortable in the discomfort. This is the work of *becoming*. The outcome of that work is that you feel right at home with your vision and ambitions, when many in the world would rather enjoy the comfort of complacency.

Accept the challenge. *Be willing.*

About the Author

A conqueror of mediocrity, Francie Van Wirkus is a certified professional master coach, helping people and teams do what seems impossible. She is a strategist, and organization and transformation thought leader, with empathy for people working in complex organizational systems.

Francie is author of a business novel series on effective leadership (not management!), The Dominant Gene, which includes three titles: Reveille, Permission to Lead, and Leading Unlikely. She also authored a memoir travelogue series, The Competitor in Me, about how we get in our own way more than anyone or anything else.

Francie is a also a certified complex trauma professional with a focus on mindfulness and faith. She enjoys racing in Ironman triathlons, teaching yoga, and lives in suburban Milwaukee, Wisconsin, USA.

Made in United States
Orlando, FL
04 November 2026

90200502R00085